Thinking and Acting in Military Pedagogy

Studies for Military Pedagogy, Military Science & Security Policy

Edited by H. Jung and W. Royl

Vol. 12

Hubert Annen / Can Nakkas / Juha Mäkinen
(eds.)

Thinking and Acting in Military Pedagogy

Bibliographic Information published by the Deutsche Nationalbibliothek
The Deutsche Nationalbibliothek lists this publication in the Deutsche Nationalbibliografie; detailed bibliographic data is available in the internet at http://dnb.d-nb.de.

Library of Congress Cataloging-in-Publication Data

Thinking and acting in military pedagogy / Hubert Annen, Juha Mäkinen, Can Nakkas (eds.).
  p. cm. — (Studies for military pedagogy, military science & security policy ISSN 1619-778X ; Vol. 112)
  ISBN 978-3-631-61580-5 — ISBN 978-3-653-03391-5 (E-Book)
  1. Military education—Philosophy. I. Annen, Hubert, editor of compilation. II. Mäkinen, Juha, 1964- editor of compilation. III. Nakkas, Can, 1977- editor of compilation.
  U405.T55 2013
  355.501—dc23

2013032007

ISSN 1619-778X
ISBN 978-3-631-61580-5 (Print)
E-ISBN 978-3-653-03391-5 (E-Book)
DOI 10.3726/978-3-653-03391-5

© Peter Lang GmbH
Internationaler Verlag der Wissenschaften
Frankfurt am Main 2013
All rights reserved.
Peter Lang Edition is an Imprint of Peter Lang GmbH.

Peter Lang – Frankfurt am Main · Bern · Bruxelles · New York · Oxford · Warszawa · Wien

All parts of this publication are protected by copyright. Any utilisation outside the strict limits of the copyright law, without the permission of the publisher, is forbidden and liable to prosecution. This applies in particular to reproductions, translations, microfilming, and storage and processing in electronic retrieval systems.

www.peterlang.de

Contents

Hubert Annen & Can Nakkas
Preface 7

1. What Can Military Pedagogy Achieve? – The Aims of Applied
   Military Pedagogy 11

Wolfgang Royl
Honor – The Personalizing Element in Military Training and Education 13

Kaisu Mälkki & Juha Mälkki
Preparing to Experience the Unexpected – The Challenges of Transforming
Soldiership 27

Hubert Annen
Managing the Grey Area – The Meaning of Rituals in a Military Context
and Their Effects on the Cadre's Responsibility 51

David Whetham
The Quest for Truth: What is a Credible Academic Source for a Defence
Studies Essay? 59

Aki-Mauri Huhtinen
The Changing Leadership Culture in Western Military Organizations 69

2. Military Pedagogic Teaching and Practice 95

Amira Raviv
Military Ethics and Moral Dilemmas: Between "On the Job Learning"
and Formal Education 97

Amira Raviv
Developing Senior Leaders: Challenges, Methodologies, and Dilemmas 109

Antti-Tuomas Pulkka & Amira Raviv
Military Ethics Instruction – The Educational Challenge of the Case-
Study Method 123

Hubert Annen
Coaching for Military Personnel – From the Idea to the Implementation 143

Luiza Kraft
Enhancing Action Competence in Military Higher Education through Cooperative Learning 155

Ulla Anttila
Military Pedagogy and Human Security Education 171

3. National Differences and Necessities in Military Pedagogy Acting and Thinking 183

Gavril Maloş & Adriana Rîşnoveanu
The Diagnosis of the Military Learning System through the SWOT Method 185

Peter Foot
Educating for Security: The Challenge to Military Academies 197

Giuseppe Caforio
The Nature of Security Threats in the Perceptions of Future Civil and Military Elites 205

Audrone Petrauskaite
Civic Values in the Context of the Military Profession: The Lithuanian Case 221

Asa Kasher
Military Ethics between Code and Conduct 229

4. Interoperability and Interculturality 239

Bo Talerud
Military Leadership for Cooperation? – Education for Civil-Military Cooperation in International Missions 241

Juha Mäkinen
Constructively Aligned Military Education and Training in Finland in the Times of the European Bologna Process 261

Sylvain Paile
Towards a European Understanding of Academic Education of the Military Officers? 279

The Authors 291

Hubert Annen & Can Nakkas

Preface

Vegetius' well-known and often misused proverb "si vis pacem, para bellum" (i.e., if you want peace, prepare for war) unavoidably leads to the conclusion that if you prepare for war, you want soldiers. Obviously, the better soldiers you have and the better they are trained, the more victorious – if not necessarily more peaceful – you will be. Evidence for this most basic military pedagogical thinking dates back as early as 1100 BC, when the Israelite judge Gideon engaged in one of the earliest documented assessments for "military personnel" (Jgs 7, 1-7), and it found its pinnacle in Napoleon's famous saying that an army's effectiveness depends on its size, training, experience, and morale, but that morale is worth more than any of the other factors combined. He arguably even was the first to quantify the psychological factor in war, claiming that in war moral power is to physical as three parts out of four. The bottom line is that wars are both fought and won respectively lost by humans.

Times have changed considerably since then, and while the ontological aspect of war itself may not have changed: the nature of conflict definitely has, and with it the nature of victory and military success. The emergence of a plethora of buzzwords (RMA, C4ISTAR, EBO, asymmetrical warfare, etc.) in military science during the last twenty years is an indicator of the obvious attempt to come to terms with this expansion of battlespace. Seldom has Clausewitz' axiom that war is the continuation of politics by other means rung more true. Armed conflict and the use of military force as a means of foreign policy have become common and even accepted in public opinion. But with this broadening of war and conflict new necessities for armed forces and their members have arisen, too. For human and social sciences have changed the Clausewitzian perception of the constancy of the human influence in war (Scales, 2009). No longer is the annihilation of enemy forces the single crucial task. Winning the hearts and minds of the native population, keeping armed conflicts below the threshold of war from escalating, preventing terrorist attacks on one's home soil by proactively engaging radical groups abroad, securing energy corridors, etc., all this has gained importance. These tasks demand a modern soldier whose skills and competences go beyond the requirements his colleague had to meet in times past, whether he's a "miles protector" or a grunt kicking in doors. They range from Civil-Military Cooperation to the Three Block War and military ethics, and may be subsumed by what can be broadly called the human dimension.

These mission- and task-specific requirements have to be identified and sensibly formulated, though. Part one of this volume is thus aimed at shedding light onto the demands that are put on today's officers and NCOs both on and off the battlefield. While the different situations a Strategic Corporal is confronted with may be quite clear, this is not necessarily so regarding the instruments and skills he needs in order to deal with them. The changing socio-political landscape of armed conflict has too often caused those responsible for training military personnel to adopt a shotgun approach, using teacher-centered teaching to administer training modules and lessons to each and every soldier. While this may assuage some policy-makers, it amounts to not much more than a military pedagogical stamp-collection of little sustainability.

Only after these specific requirements are identified and operationalized is it possible to develop the necessary pedagogical and didactic tools and deliver them to the appropriate target groups. It is a sign of the times that modern civilian teaching techniques have found their way into the military classroom (Syme-Taylor, 2010), including coaching, psycho-education, small group teaching, jigsaw method, case studies, etc., as part two will show. This development has been necessitated by the new demands and difficulties soldiers have to deal with in modern operations. The Revolution in Military Affairs, the effects of network-centric warfare, and the transformation of armed forces worldwide have been characterized by developments that the civilian industry and private sector have had to come to grips with decades ago, e.g., technology and computerization edging out human labor. Downsizing and the increasing application of hi-tech in the armed forces has resulted in a reduced need for the military equivalent of unskilled workers, and has gone hand in hand with a rising need for well-educated specialists and experts – or at least persons of high-potential. Modern armed forces thus face the similar problem the U.S. Army was confronted with during World War 1 when they commissioned the Army Alpha and Army Beta Test: identifying intellectually apt men and women capable of manning and operating state-of-the-art equipment. And just as behaviorism was highly influential in developing military training procedures during the Great War, so do current armies employ modern didactics and pedagogy to teach their men and women in uniform the skills and competences they need to meet the diverse and heterogeneous demands of the full spectrum of military operations, be it warfare or military operations other than war (MOOTW). The quote of a former chairman of the Joint Chiefs of Staff during the 1990s that "real men don't do MOOTW" nowadays seems quaint at best and embarrassing at worst.

Part three shall focus on national differences and necessities in military pedagogical thinking and acting. Since political parameters and restrictions (e.g., constitutions, security policies, membership in supra-and/or international bodies, etc.) provide the framework within which a nation deploys its armed forces and engages in military operations, these national idiosyncrasies evidently affect the philosophy of its military training and thus have also important consequences for the education of officers and NCOs. But differences and distinctions alone should not be sufficient reasons to discount experiences alien to one's own empirical world. For example, the development of a foreign armed forces' curriculum for officer education (Simpson, 2008) may on the one hand serve as an informative view of this nation's otherness, consequently allowing us to re-assess what is specific of ourselves. On the other hand, it may also serve as a case study that lets us see shared structures and *topoi* that transcend national particularities, giving us a glimpse of the broader context of military pedagogical endeavors. This approach can meanwhile be rightfully called a military pedagogy tradition, as exactly one decade ago volume 8 of the "Studies for Military Pedagogy, Military Science & Security Policy" was published, titled "Military Pedagogy – an International Survey" (2002). In its preface, editor Heinz Florian had identified some communication problems amongst international experts in military education with regard to the application of the word *pedagogy* in the context of military training and education. He thus set out to create a common basis for military pedagogy by presenting the various understandings of the concept.

But he had also found agreement in two important points. First, the necessity of increasing international cooperation in future military tasks. Secondly, improved interoperability as a prerequisite for military success. A decade later these points haven't lost any importance, on the contrary. The sight of allied forces working and soldiering together has become commonplace in most operational theatres, and winning the civilian population's hearts and minds has remained as important and difficult as ever. Part four shall thus highlight thoughts and attempts at creating and enabling interoperability by fostering interculturality and an awareness for national particularities.

Ten years have passed since then, and while some naysayers might lament that there is still no internationally accepted "unified field theory" of military pedagogy, optimistic realists will be able to see that military pedagogical research and teaching has made headways. The diversity and heterogeneity of military pedagogy can be likened to a burlap sack getting filled with different objects: this may change the sack's shape, but it also gives it a more pronounced and recognizable outline. The articles in this volume are from the 9th and 10th International Conferences on Military Pedagogy organized by the National Defence University in Helsinki and the Defence Academy of the United Kingdom in Shrivenham, respectively. While political and educational developments have caught up with some of the topics addressed, changes in the military, political, and educational landscape will always necessitate this branch of academia to continuously adapt to the needs of the armed forces and their servicemen and -women. Insofar all the contributions in this volume offer valuable insights into current military pedagogical thinking and acting and present milestones in the field of military pedagogy.

Bibliography

Florian, H. (2002). Preface. In H. Florian (Ed.), *Military Pedagogy – an International Survey* (pp. 5-6). Frankfurt am Main: Peter Lang.

Scales, R. (2009). Clausewitz and World War IV. *Military Psychology, 21 (Suppl. 1)*, 23-35.

Simpson, H. (2008). Designing a Defence Studies Curriculum for the Republic of Sierra Leone. In T. Kvernbekk, H. Simpson & M. A. Peters (Eds.), *Military Pedagogies and Why They Matter* (pp. 63-75). Rotterdam & Taipei: Sense Publishers.

Syme-Taylor, V. (2010). Innovative Teaching Methods in Military Pedagogy. In H. Annen & W. Royl (Eds.), *Educational Challenges Regarding Military Action* (pp. 209-214). Frankfurt am Main: Peter Lang.

1. What Can Military Pedagogy Achieve?
– The Aims of Applied Military Pedagogy

Wolfgang Royl

Honor – The Personalizing Element in Military Training and Education

In the following, the idea is rejected that the meaning of honor should be imparted sociologically by dealing with it through the concept of social prestige. Rather, thoughts will be presented and assessed devoted to a differentiation of honor concepts. Differentiating between the honor bestowed by others and one's own concept of honor takes into consideration the way and mode of how 'honor' may be experienced as an outward and an inward phenomenon of consciousness, and how it should indeed be experienced by our soldiers. In this process, the value of being a soldier is determined by the fact soldiers make a pledge to bravely serve their nation and even to sacrifice their lives in the line of duty (i.e., unlimited liability). This is what constitutes the essential difference between soldiers and the members of any other profession or trade. If this were to be taken into account, the excellence of military existence would need to be shown by ascribing to it a value, which would allow an individual concept of soldierly honor to be developed from it that should correspond with the public ascription of soldierly honor. The double borderline phenomenon of perception psychology is discussed so as to illustrate this correspondence relationship. After all, the experience of honor being bestowed 'externally' and the experience of honor as derived from one's own self-estimation are not perceived in isolation, but as related events with mutual dynamics.

Dealing with honor

"In the Europe of Old, 'honor' meant a social class' claim for prestige that was raised to an exemplary status, and at the same time it meant the attitude of representing, enforcing and maintaining this ethos of the exemplary." (p. 152). This time-related placement and definition of honor provided by Pankoke (1994) may serve as an introduction to the topic. Similarly, Julian Pitt-Rivers (1968) differentiates between three dimensions of honor: "Firstly, it is a sentiment; secondly, it is a manifestation of this sentiment in conduct; and thirdly, the evaluation of this conduct by others".

As an example, Johann Friedrich Adolph von der Marwitz (1723-1781) should be called to mind. Reportedly, he refused to let the forces under his command pillage the elector's hunting lodge *Hubertusburg* near Leipzig during the Seven Years' War. By doing so, he fell into disfavor with Frederick II, King of Prussia, who then gave permission to a corps of volunteers to plunder the lodge, which was famous for its art treasures. As a result, Marwitz decided to leave the military service. The epitaph on his tombstone reads: *Wählte Ungnade, wo Gehorsam nicht Ehre brachte* [Chose disfavor over obedience that would not bring him honor].

This episode depicts honor as a stable attitude towards the world. Obviously, Marwitz, an educated nobleman who is said to have been highly appreciative of arts, had a clear concept of what his reputation demanded from him. As he was able to value works of art not only in financial terms but also for their artistic importance, he could not possibly tolerate that they be handled in such a barbaric manner. Certainly, to desist from taking away such cultural assets in a careless fashion was not an easy decision to make, as in those days the enlisted men of any army would raise their meager pay by selling what they had looted. What made this event so important for Marwitz was the fact that he felt his individual honor to be offended by the king's decision. The honor of serving under Frederick II and, possibly, participating in the king's future glorious battles meant less to him than his personal appreciation of the fine arts and their irretrievable objectivations.

If a well-known art expert and art connoisseur were known to handle precious objects of art in an unskilled and uncaring manner, he would suffer a loss of honor. Other people knowledgeable in the subject matter, who might have held him in regard before, would lose their respect for him. This loss of honor in the eyes of the public would not fail to affect the individual's self-esteem. Once a person has displayed a deviating behavior, he or she is assumed to do it again. This comparison may serve to illustrate that Marwitz, who was integrated into higher social circles whose members were appreciative of arts, could not execute the king's pillaging order for that very reason. Honor as an internal concept shares a common boundary with the honor bestowed on an individual by the outside world. In terms of his appreciation of honor, Marwitz is a historic example.

In contrast, in their assessment of honor and its value in a theoretical framework, Vogt and Zingerle (1995) arrived at a more contemporary perception: Honor remains a "general structural element of the modern societies, i.e., in its relation of personal identity and social role." (p. 13). Here, these authors have unmistakably based their listing of the reference fields of the semantics of honor on a ranking order, among other factors, in that they consider 'honor' to be a state of affairs which provides for a value-related and graded differentiation of an individual's relationship to other people. Thus, 'honor' also becomes the "determinant(s) of identity, of social standing (status) and of morality." (p. 16).

The example of Major General Gerd Schultze-Rhonhof should be called to mind in order to emphasize the coherence of these three criteria, which help to define the matter of 'honor'. As for the professional identity of this officer, we may rightly assume that he fully identified himself with the oath of office he took as a military man. As his status was that of a brigade commander, one may also assume that this soldierly identity should have translated into his acting as a role model, which he will have wanted to be both for the officers and the enlisted personnel under his command. The same is expected from any military leader. Public derision of the military profession cannot be tolerated by such a leader if he takes his mission and his profession seriously, as it would be detrimental to his honor and the honor of his subordinates.

His counterpart Peter A., a physician and conscientious objector, who was present at a Bundeswehr information event held before a class of high school students, referred to soldiers as potential murderers. Later on, the Federal Court of Justice ruled that he may do so under the right of free speech. The physician was acquitted.

In turn, Schultze-Rhonhof, while not directly involved but feeling that his honor had been tainted, took it upon himself to raise a question relevant to officers and enlisted personnel: Why still show bravery, then? He called to mind the soldierly virtues and made it clear how a soldier's self-esteem could be affected if the honorable mission assigned to the armed forces by the constitution were allowed to be publicly doubted without fear of punishment.

It was the character-stabilizing function of 'honor' around which political scientist Richard A. Gabriel (1982) focused his critical analysis of the Vietnam War. During the war (1965-1975) the U.S. forces were exposed to circumstances which for them were extraordinarily stressful. Many officers and enlisted personnel lacked the moral fortitude to withstand these pressures, and the outcome of the war was inglorious. In an effort to reconstruct the moral preconditions for soldiers deployed in the field, Professor Richard Gabriel made the subject of 'honor' an integral component of a person's character. In more concrete terms, he described 'honor': "It is an integrating trait of the soldier's character, and it prevents the application of technical military skills from becoming an exercise in moral horror". His suggestion that the guiding motto of the U.S. Military Academy at West Point, 'Duty, Honor, Country', should be rephrased to read 'Honor, Country, Duty' goes to show the importance which Gabriel attaches to honor as a part of a person's ethical obligations. So as to translate this shift in emphasis into practical terms, Gabriel suggested that military courts of honor should be established which would then be tasked, as panels acting independently of the public judicial system, to investigate cases of dishonorable behavior committed by armed forces personnel while on duty.

Apart from the fact that Vogt and Zingerle, who are authors of repute on the subject of honor, document the contemporary meaning of 'honor', much like Eric H. Erikson they refer to the topic of identity, one of whose determinants was honor, as they claim. Such an identity, an agreement with oneself, may come up as a matter of course in our daily lives, and may be experienced in an immediate manner. If, however, the morals of an institution are concerned and, in particular, of an institution that provides fighting potential (i.e., weaponry and soldiers) for Partnership for Peace and other missions, it is indeed appropriate to entertain further-reaching thoughts as to the qualification of the moral conscience of the soldier. Reports on UN soldiers in Darfur who force starving girls and women into submission in return for food would suggest that it is necessary to integrate the norms of soldierly awareness of honor into military education and training on an international scale.

This is obvious for another reason: In his analysis of social facts, the sociologist Georg Simmel (1892) concluded that "honor is what turns a man's social duty into his individual welfare, thus moving him to make the most enormous sacrifices" (p. 405). Sometimes, this may also apply to operations in foreign countries, where military units vie to outdo each other. For the missions of preventing violence between the estranged factions of a nation's people or, to use another example, of safeguarding the reconstruction work in Afghanistan against terrorist threats, make demands on the soldiers far beyond the levels of risk that exist on the civilian labor markets of the sending nations.

Applying this structural estimate of – and the treatment of – honorable behavior to the military sphere, it would oblige each soldier to act in an honorable way. This is particularly illustrated by the pledge of allegiance to the flag or the oath uttered by soldiers. The words "I vow" or "I swear" individualize the obligation imposed on them. As a rule, such ceremonies are held at historically prominent locations, possibly at night and under the light of torches so as to appeal to the participants not only intellectually but also emotionally. Particular effort is made to provide a solemn setting to reach each person in his or her entirety. The great social importance and the psychological effectiveness of such solemn military ceremonies for society on the whole and for the soldiers in this particular situation is demonstrated – *ex negativo*, as it were – when anti-military groups attempt to disturb the public performance of a ceremony. Such negative displays of behavior have not been observed in nations where the citizens have a positive attitude towards their army and acknowledge the forces' readiness. There, the soldierly honor of risking life and limb in serving the nation is recognized by the general population showing their appreciation of the armed forces.

The construction of the honorable

Honor is a good example for illustrating how modes of behavior – and this applies to soldiers, too – are based on the effects from both 'inside' and 'outside'. For honor is felt in a dual way: On the one hand, as the individual's feeling for the moral demand level within him- or herself and, on the other hand, as a measure of respect and appreciation received from others. The relationship between 'inside' and 'outside' is explained by means of the 'double borderline' phenomenon of perception psychology.

On one side of this borderline, the individual perceives honor as something that is part of him- or herself. On the other side of this differentiation which may be experienced by any person, there is the individual's perception of honor being bestowed by others. The same applies to the loss of honor, when people regarded as significant by the individual withdraw their appreciation.

Conflicts will arise when an individual tries to defend his or her honor, or attempts to recover it after having been deprived of it by some other person. The custom of challenging someone to a duel is a historical example (Frevert, 1991). Nowadays, the dishonorable discharge of a soldier from the service for severe violation of military rules is governed by applicable service regulations.

The phenomenon of the double borderline

A double borderline separates what is to the left of it from what is to the right of it, and vice versa. Honeycombs may be used as a visual and quite illustrating example (Fig. 1).

Figure 1. Illustration of the double borderline.

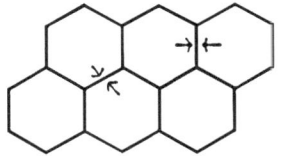

The hexagonal cells of a honeycomb (Fig. 1) share only one cell wall with each of its neighboring cells. This results in a group of cells being connected to each other. Each cell wall serves as a separation of its content, i.e., the inside, and as a separation of the content of the neighboring cell, i.e., the outside.

This phenomenon of the double borderline is appropriately described by a distich written by the German poet Johann Wolfgang von Goethe (as translated by Christopher Middleton): *"You must, when contemplating nature, attend to this, in each and every feature: There's nought outside and nought within, for she is inside out and outside in"*

Applied to the topic of honor, this four-line poem expresses the interesting characteristic of the conscience of honor and feeling of honor. The moral feeling and conscience of honor reflected in it refers to the estimation and control of one´s own behavior towards others. Also, it is determined by the social estimation of one's own honorability by other people. One aspect of the implied context is to describe and understand the feeling of honor as an internal, individual feeling. The other aspect is someone being considered as honorable by other people, i.e., from outside.

This interrelation of ascribed honorability and one's own concept and feeling of honor starts to exist as soon as a child has grasped the difference between 'yours' and 'mine' and then is blamed for having taken something away from someone else, i.e., for having stolen something. The same applies if one nation secretly taps the oil deposits located beyond the borderline which it shares with a neighboring state, thus misappropriating the resources of the other nation. Such actions may give reason for military conflicts. Just like a private individual would react if something had been stolen from him or her, nations, too, will take the matter before the International Court of Justice to protect their rights, or they may choose to resort to military means forthwith. In some way or other, attempts will be made to maintain or restore the nation's honor within the international community. Governments feel required to take action not least because they are obliged to do so for the sake of their own citizens, i.e., because of their domestic responsibilities.

Using the concept of honor to define what is 'soldierly' opens up a dimension of the military which is of particular importance both for the community of soldiers and for the individual soldier.

Personal honor

The quality of military training will make a difference as to whether a soldier considers his or her profession a job like any other, or rather as a calling. As such, the commitment to serve in the military is distinguished from the rules that apply to other trades or occupations by the recruits' pledge or oath. In the German Bundeswehr, recruits serving as conscripts are required, upon entering the service, to make a solemn pledge to "loyally serve the Federal Republic of Germany and to bravely defend the right and the liberty of the German people", whereas the oath to be taken by voluntarily enlisted starts with the words 'I swear' instead of 'I pledge'.

Normally, the organizers succeed in carrying out the ceremony of the pledge of allegiance to the flag in such a way that all participants, including the attending family members, are emotionally touched and become aware of the 'solemn' character of the event. Speaking the words of the oath is considered to be an individual's personal acknowledgment of military service or the military profession. Those who have taken such an oath are expected to identify themselves with the military duties transferred to them.

The extent to which military honor develops into an integral part of a soldier's military bearing, however, will depend on the circumstances of military training and education. In the optimal case, an individual's identification with the soldier's trade will result in a personal identity which subsumes the person's inner attitude towards the military profession and the accomplishment of military duties to form an intrinsic part of that individual's personality.

As was shown by Erikson (1956), 'identity' denotes a psycho-social set of facts. The hyphenated attribute 'psycho-social' illustrates the connection between the 'inside' and the 'outside' of the individual, quite in keeping with the double borderline phenomenon. With reference to the individual, the 'ego identity' is considered to be a part of the 'ego'. The development of the ego identity, i.e., of being identical with one's own self, is described by Erikson from the psychoanalytic point of view as a function of the ego synthesis. There is an interlinking process of identity elements arising from a particularly intensive identification of an individual with other individuals and with sets of facts. The individual ego identity is enhanced by such processes of meeting and learning. The mental structure of a person, in our case the soldier, is widened by various offers of and opportunities for identification, and provided to the soldiers as they are integrated into the military community. For the military leaders, this means that they have to accomplish an important psycho-hygienic task. As is well known, the sleeping quarters of soldiers are the places where recruits – whether this is to their liking or not – are 'brought up to speed' regarding any knowledge they may still lack in sexual matters. As a rule, the number one topic in this evolving club of men is going to be willing women and the experience made with them. It may even happen that an older soldier will tell newcomers of a nearby brothel's specially priced services for regular customers. Keeping such matters within proper bounds while giving priority to the main thing – the purposes of military service, that is – will set the conditions required for soldierly honor to develop. This is why recruiters choosing candidates for the careers of noncommissioned and commissioned officers will pay strict heed to the selection of applicants of steadfast character with firmly established ethics. It is for good reasons that performance ratings are required at regular intervals to assess the leadership skills and performance of those who are given responsibility as military leaders. Lack of leadership abilities as well as personal conflicts, if any, may thus be identified and dealt with at an early stage.

All of this will be helpful in creating a military environment that promotes the positive development of the soldiers' identity with their trade, i.e., the soldierly character. Military virtues such as camaraderie, bravery, and loyalty will – if they have developed into elements of this identity – enable the soldiers to remain steadfast when exposed to stress. It may be safely assumed that during operations in foreign countries the level of the soldiers' resilience against post-traumatic stress disorder (PTSD) is increased if the soldiers identify themselves with their military mission, and if their military operations are part of their military identity. One should not underestimate the difficulty of harmonizing this identity with the demands raised by military honor, though. This difficulty is illustrated by the example of Florian P., a Bundeswehr major who refused to assist in the programming of software which might have been used for solving logistic problems of NATO allies participating in the war against Iraq. The major's case was an exceptional one. He passed the medical examination he was ordered to undergo to assess his mental state, but would not budge from his view that his disobedience was justified because the war in Iraq was unlawful. He won the appellate hearing before the German Federal Administrative Court in Leipzig, where he had lodged an appeal against a military court which had ruled against him on grounds of disobedience. He was acquitted and was adjudged the right of refusing to participate – be it directly or indirectly – in a war that was in violation of international law. The major withstood the disciplinary pressures put on him, even when he was disregarded for promotion. He published a book with the provocative title *Totschlag im Amt* ('homicide in the line of duty') (2008) that reveals him to be a person who uses meticulous argumentation to describe his concepts of personal and military honor. Also, the book describes how his superiors, even the higher staff officers, had made efforts to infuse self-doubts into his mind about his individual and military integrity and to question the honorableness of his disobedience.

While the case of Florian P. is worthy of note, it is certainly not a prime example of how a soldier should try to draw attention upon himself; and the same could be said of his superiors who failed to take this officer and his concerns seriously in due time. Had a soldier of such individual and military integrity like this major been assigned to the Abu Ghraib U.S. military prison in Iraq, the transgressions against Iraqi POWs would surely never have occurred. For a Bundeswehr soldier, the duty of obedience is waived whenever the execution of an order would be in violation of human rights.

Social honor

Personal honor, i.e., the ideas on honor considered by the individual as his or her own, is the identifying reflection of moral lessons learned. Receiving unfair treatment will sharpen a person's sense of justice. The American military author Richard Gabriel (1987) commented on the actions taken by the U.S. Army during the Vietnam War, criticizing officers who – for the sake of their careers – put aside their moral concerns about the way the war was waged. In doing so, he sided with the soldiers who took seriously the motto created by General Douglas MacArthur in his speech at the U.S. Military Academy at West Point on May 12, 1962, which was to become a hallmark of military ethics. Since then, the poignant sequence of 'duty, honor, country' has symbolized the social requirement for any soldier to act in accordance with these guidelines. Life at West Point follows the brief and concise rule: "A cadet will not lie, cheat, or steal, or tolerate those who do!".

Because the verbal expression of such behavior objectives is not sufficient to guarantee that they will be put into practice, it is necessary to explicitly demonstrate military ethics and their moral applications in military training and education on a continuing basis. In the well-known book by Malham Wakin and Lewis Sorley (1981), a former member of the U.S. Air Force Academy recommends: "Periodic examination of the ethical imperatives of the motto 'duty, honor, country' and of the relationship of the military profession's precepts and its practices in light of these imperatives, is a necessary and rewarding experience" (p. 143). Routine duties on post, and military operations in particular, offer plenty of opportunities to put this recommendation into practice whenever conflict-prone situations arise.

Lawrence Kohlberg (1981), the American moral psychologist, has provided us with an overview of stages of moral development that has gained a global reputation. This sequence of steps has proved itself to be appropriate for analyzing ethical dilemmas, and thus not only for advancing the ethical concepts of individuals or group members, but also for developing acceptable solutions for actual conflicts. In what Kohlberg referred to as the captain's dilemma, an officer has to decide on who of his men is ordered to attach the explosives to a bridge in the face of the approaching enemy. It is quite likely that there will not be enough time for the soldier attaching the explosives to get away from the bridge in time. Most certainly, the enemy will shoot him. Now, who should be given this dangerous mission? An older soldier, who has a family he needs to take care of? An unmarried younger soldier who has no dependents? Can the captain risk the loss of a defense specialist whose services are urgently needed for an upcoming mission? Which decision would be justified in moral terms?

The phenomenon of the double borderline may be illustrated – though less dramatically than in the captain's dilemma – by the way soldiers treat their uniforms. To the outside world, the soldiers' uniforms show them to be members of the military. At the same time, the uniform, as a part of the physical ego, symbolizes something like an internal soldierly attitude. It is not for nothing that recruits are taught to adhere to the military 'dress code' during their training. Normally, a large mirror will have been installed for that purpose in the exit area of the company building. Every soldier is expected to check his or her outward appearance before leaving the building. It is the quality of military training and education that will determine whether soldiers consider it a nuisance to be inspected for proper attire by their sergeants during roll call, or whether they regard it as a duty and honor to adopt the prescribed standards of appearance and military bearing as their own.

By accepting the duty regulations and serving their nation and its citizens, the soldiers in uniform acquire their profession's typical behavioral standards, which distinguish them from people who pursue other trades or occupations. Democratic societies, however, which on principle are based on the rule of equality (as expressed in the general appreciation of the individual) necessarily tend to refrain from emphasizing the differences that may exist regarding the social prestige of trades and occupations. After the Second World War, this gave rise to the idea of the 'citizen in uniform' along the lines of Baudissin's Leadership Development and Civic Education concept, an interpretation going back to the French War of the First Coalition and the *levée en masse* (general conscription of 1793).

In the European nations, though, the memory of this historical role model has not caused people to enthusiastically embrace the idea of general conscription. Among university students in particular, the number of conscientious objectors is above average. The reduction of the German Bundeswehr's authorized personnel strength has diminished the demand for conscripts. Additionally, because of the pay raise for temporary-career volunteers, there is, at least for the time being, a sufficiently high number of troops available to man the contingents for deployments abroad. These facts conceal the precarious situation of a *de jure* conscript force, but meanwhile *de facto* volunteer force that is losing the support of its people, who are less and less willing to do military service.

Because this situation is reflected on an international level as well, some governments have abolished compulsory military service. These nations have opted for an all-volunteer force. The constitutional principle of equality, in this context referring to compulsory military service, has thus been abandoned by the members of those societies for whose benefit it was originally written into the constitution. Many a citizen who would be fit for military service does not consider it an honor to become a soldier.

Hans Speier (1952) wrote that human beings differ as to their concepts of honor, a fact which almost in and of itself marks the social differences: "From the fact that honor is derived from a concept of excellence it is inevitable that the process of honoring creates hierarchical distinctions" (p. 45). Speier continues: Honor sets the rules for us to guide our lives and causes anybody who does not follow them to be socially excluded. In this way, honor has a disciplining effect as a social element. As such, it sets limits for the individuals in his or her behavior. This is especially true in the military.

Now, this fact is the link to the original idea of granting the military a status *sui generis* in our society. This might be justified because soldiers take on the duty and burden of national defense on behalf of all other citizens, and they even do so outside of their own country, like in Afghanistan. Former German minister of defense Dr. Peter Struck, now sitting in parliament as a member of the Social Democratic Party, stated that German security is also defended in the Hindu Kush. The minister said this to counter the argumentation of his political opponents, who insisted that according to the German Basic Law, the Bundeswehr's exclusive mission was to protect the national borders of the Federal Republic of Germany, a precaution which luckily had been turned obsolete, as formerly antagonistic nations had joined together in the European Union. In other Western nations, such as the US, France, and the United Kingdom, the need to conduct military operations beyond the national borders has caused governments to institutionalize an all-volunteer force to ensure the required level of operational readiness.

It has been observed that because of such a drastic change the armed forces were no longer able to recruit personnel analogous to the structure of the overall society. The recruits drafted into the armed forces do no longer reflect all strata of society. It is often the members of the lower social strata who consider military service as a chance to secure permanent employment and to seek career opportunities that would allow them to climb the social ladder.

The British forces intend to establish residential areas so that soldiers will be able to live at their duty station, sparing them the prolonged separation from their families. Moreover, a large-scale civilian-military urban development will provide jobs for family members seeking employment in a civilian vocation or profession. Also, opportunities are to be provided for soldiers to purchase a home. A school system is planned that will provide educational opportunities for children and teenagers up to and including the qualification for university-level studies. Another idea is to increase the attractiveness of military service for the children and teenagers of soldier families, as even today many recruits come from families where one parent has served in the military. It almost goes without saying that, in such a model-of-excellence development for military personnel, appropriate support will be given to families whose members were wounded or killed in action.

Granting such privileges to the military is going to change the self-awareness of the soldier. For if training and education make it possible for the individual soldiers to witness their own concept of honor as a reflected image of the appreciation by society, this will promote the intensity of their personal feeling of honor. Being a soldier would then correspond with a self-related concept of honor which may also have a stabilizing effect on the soldiers' behavior whenever the boredom of routine activities might affect their motivation. Taking pride in being a soldier may suitably protect the soldier against those who openly show their contempt for service personnel in an attempt to denigrate the military profession. The soldiers' personal feeling of honor will safeguard them against having their self-conception shattered by others. The soldier knows of the value of being a person entrusted with defending the safety and freedom even of those citizens who revile soldiers and have a poor opinion of their commitment. It appears to be appropriate, from a social-ethical point of view, for members of the military to develop the self-conception of an elite, as they have committed themselves to risk life and limb in defending the Free World.

Bibliography

Erikson, E. (1956) Das Problem der Identität. *Psyche, 10* (1-3), 114-176.
Frevert, U. (1991) *Ehrenmänner. Das Duell in der Bürgerlichen Gesellschaft.* München: C. H. Beck.
Gabriel, R. (1982). *To Serve with Honor. A Treatise on Military Ethics and the Way of the Soldier.* Westport: Greenwood Press.
Gabriel, R. (1987). Legitimate Avenues of Military Protest in a Democratic Society. In M. Wakin, K. Wenker & J. Kempf (Eds.), *Military Ethics* (pp. 101-117). Washington, DC.: National Defense University Press.
Kohlberg, L. (1981). *The Meaning and Measurement of Moral Development.* Worcester: Clark University Press.
Pankoke, E. (1994). Zwischen "Enthusiasmus" und "Dilettantismus". Gesellschaftlicher Wandel ‚freien' Engagements. In L. Vogt & A. Zingerle (Eds.), *Ehre, archaische Momente in der Moderne* (pp. 151-171). Frankfurt am Main: Suhrkamp.
Pitt-Rivers, J. (1968). Honour. In D. Sills (Eds.), *International Encyclopedia of the Social Sciences, Vol. 6* (pp. 503-510). New York: Elsevier.
Schultze-Rhonhof, G. (1997). *Wozu noch tapfer sein?* Gräfelfing: Ingo Resch.
Simmel, G. (1892). *Einleitung in die Moralwissenschaft.* Berlin: Wilhelm Hertz.

Sorley III, L. (1981). Duty, Honor, Country. Practice and Precept. In M. Wakin (Ed.), *War, Morality, and the Military Profession* (pp. 143-161). Boulder: Westview Press.

Speier, H. (1952). Honor and Social Structure. In H. Speier (Ed.), *Social Order and the Risks of War* (pp. 36-52). Cambridge: Massachusetts Institute of Technology Press.

Vogt, L. & Zingerle, A. (1995). Zur Aktualität des Themas Ehre und zu seinem Stellenwert in der Theorie. In L. Vogt (Ed.), *Zur Logik der Ehre in der Gegenwartsgesellschaft. Differenzierung, Macht, Integration* (pp. 9-34). Frankfurt am Main: Suhrkamp.

Kaisu Mälkki & Juha Mäkki

Preparing to Experience the Unexpected – The Challenges of Transforming Soldiership[1]

Introduction

> *[...] so in our profession, many are to be found who know every precept of it by heart; but alas! when called upon to apply them, are immediately at a stand. They then recall their rules, and want to make everything, the rivers, woods, ravines, mountains, etc. subservient to them, whereas their precepts should, on the contrary, be subject to these [...].*
>
> Major General Henry Lloyd, *The History of the Late War in Germany*

Henry H. E. Lloyd (1718-1783) pointed out that there were two parts to the art of war: a mechanical part that "may be taught by precepts" and by mathematical principles, and another part which has no name, "nor can it be defined nor taught. It is the effect of genius alone." After a few decades, Napoleon Bonaparte and his armies shook the foundations of European military tradition and cultural heritage by putting these "enlightened" critiques into practice and taking full advantage of the "weaknesses" of current military thinking. The reason for Napoleon's long run of victories lay in his opponents' inability to understand his way of fighting and of devising effective responses. Napoleon's army relied on new methods of training, organizational changes and doctrinal innovations (Paret, 1990). Napoleon also relied on the idea of a dispersed battlefield and dispersed operations (DO), and above all, he relied on knowing about the limits of his adversaries, i.e., their cultural "Achilles heels" stemming from their collectively followed and actualized habits of mind. This was basically the core of the non-mechanical part of the art of war which led Napoleon to triumph over his opponents, despite the fact that it was difficult to define or teach. In other words, Napoleon chose and exploited the line (or course) of least expectation (Liddell Hart, 1991), i.e., exploited the adversaries' line or courses of natural expectation.

Complexity and uncertainty undoubtedly are, and have always been, characteristics of warfare. The defining element of this is that future war is expected to be always unexpected and in many ways chaotic. So the necessary and often-asked question is how is it possible to prepare for the unexpected? Yet at the

---

[1] Originally published as Luoma, K. & Mällki, J. (2009). Preparing to Experience the Unexpected. The Challenges of Transforming Soldiership. *Tiede ja ase, 67*, 110-132.

same time we have to keep in mind the well-known risk for militaries: that they would nevertheless be preparing for the past wars. Thus another necessary question is, how can we know whether we are preparing for the past war? And furthermore, if we somehow find ourselves leaning on the familiarities of the past, how is it possible to overcome this view that is expected to be limited in terms of future contexts? These three questions are very much intertwined. The orienting magnetism of the past is deceivingly invisible as it is dressed in the clothes of normality. The tension between this normality and the unknown future and acting with knowledge gaps is also evident in Lloyd's excerpt above, where the inductive interpretation of the current situation is overrun by the habits of expectation. Being able to interpret situations inductively is crucial when the situations are unexpected, that is, when we do not know in advance what is going to happen and thus cannot educate military leaders accordingly.

This portrayal of Napoleon's art of war can be seen to include many of the current challenges of military pedagogy, although the actual theme of transforming the soldiership, i.e., the culturally constructed idea of soldier's being, has been mingled with the themes concerning irregular warfare; terrorism, insurgency and asymmetric warfare. The most important player in irregular warfare is *not*, however, the state where the actual war is being fought (Carrick, Connelly & Robinson, 2009; Kiras, 2008), but the soldiership that is being exposed and challenged by the faces of war that "distinguish real war from war on paper" (Clausewitz, 1989, p. 119; Mälkki & Mälkki, 2011). In the present-day argumentation concerning the theme of modern battlefield, perhaps too much emphasis is being put on the imagined changes on the conduct of warfare and the nature of the potential adversaries, instead of us as humans involved in warfare and the cultures that ultimately define each belligerent. Hence, warfare is always the "collision of two living forces" as it is based on two-sided interaction where both belligerents end up facing the true conduct of culturally influenced but also biologically functioning human behavior (Clausewitz, 1989; Mälkki & Mälkki, 2011).

Military pedagogy describes the demand both to locate and define the present state of the art of war and the lines of natural expectation, and to be able to change or transform soldiership in order to make progress instead of repeating the habits of the past. Actually, weaknesses in one's habits of mind can be seen as the very object of such transformation. We may still be in the middle of a technological revolution concerning military equipment and the communication methods (Lonsdale, 2008). Despite of the fact that these innovations are constantly changing the conduct of war, it is reasonable to recall, that the human being beside the advanced and constantly inclining materials, is still the same as he was hundreds and even thousands of years ago. Corporally, we are equally unarmed in the front of the harsh battlefield conditions, equally incapable of perceiving irrational decisions and regrettably powerless while trying to renounce violence (Grossman, 2007; Mälkki & Mälkki, 2011; Shalit, 1988). Hence, one of the continuous challenges of military pedagogy is how to support a soldier's ethicality as well as personality and competences in order for him/ her to be able to cope with the modern battlefield.

Within discussions on military pedagogy, transformation of soldiership has been claimed to be needed in order to build more adequate grounds for operating on modern battlefields (Mäkinen, 2006; Toiskallio, 2004). Action competence and cognitive readiness, among other suggestions concerning the goals of transformation, are offered as conceptual tools for preparing for the unexpected. However, what is still missing is an understanding of the transformation itself (Toiskallio, 2004). That is to say, we have the goals and guidelines of the transformation, but the logic and process of the transformation itself is unexplored: how does it happen? How do we transform ourselves into the action-competent soldiers of the modern battlefield? What is the transformation like that would build the ethicality of the soldier and support their personal development and emancipation (Pulkka & Raviv, 2007; Toiskallio, 2005)?

And furthermore, what does it mean to transform ourselves for the paradoxical mission of being prepared for the unexpected (Mäkinen, 2006)? In this it is essential that we aim to understand *why we are not yet there*, that is, *what is it that keeps us the way we are*, and above all, *what is it that we are now* since we obviously are not there yet, as the military organizations seems to be in a need of transformation (Lonsdale, 2008). Only then can we understand and aim to foster transformation. However, these essential viewpoints have been largely neglected within discussions on military pedagogy. Perhaps this is so because in these questions concerning the transformation of soldiership, the individual and psychological dimensions are intertwined with the social and sociological, and are shaped by the current art of war that mirrors the mindsets of society (Hartman, 2000). Therefore we will undoubtedly fall short of understanding the prerequisites for a transformation of soldiership if we examine individual competences apart from the cultural and organizational context of the soldier's actions, or if we reduce the context to mere material and technological elements without the influence of the human collective. Instead, we need an understanding of the *individual within his or her context*, and *how the context lives within the individual*.

These questions and issues can be seen to be at the heart of action competence, the core construct of military pedagogy. In this paper we attempt to tackle this challenging issue at the crossroads of the personal development of the soldier and the efficiency of the military organization. The bases of our approach are found in the transformative learning theory of Jack Mezirow (1991; 2000; 2009) which, as an adult learning theory, provides an excellent basis for examining the transformation of soldiership and action competence based on the following viewpoints. First, the theory deals with transformation in the context of learning. Second, it takes into account the earlier experiences and existing habits of mind and habits of expectations as the basis for learning. Third, it recognizes the influences of socialization and the cultural context on the mindset of an individual while mapping the terrain for finding one's own voice. The fourth viewpoint, and most interesting one in terms of complex military contexts, is that transformative learning theory has at its core meaning-making, interpretation and understanding, as well as also considering the situations in which we do not automatically understand – that is, it deals with making meaning in situations of chaos and meaninglessness.[2] Thus it points to the very core of the military pedagogical challenges of preparing for the unexpected and making decisions in the face of uncertainty.

In the considerations of this article Mezirow's theory is utilized in order to approach issues that may be seen to lie on the borderline between the fields of military pedagogy and military psychology. That is to say, we aim to take into account both influences of the stressful context to soldier's cognitive performance and the issues related to learning to deal with and prepare for these. In this it is needed to build a bridge between the educational considerations focusing on learning (Mäkinen, 2006; Royl, 2005; Toiskallio, 2000, 2004) and the psychological considerations on the effect of affect to one's performance (Bartone; 2006; Litz, 2007; Mälkki & Mälkki, 2011; Tenenbaum, Edmonds & Eccles, 2008; Wallenius, Larson & Johansson, 2004).[3] Thus the focus of the article is not on action competence per se or military psychological considerations per se, but on aiming to grasp a phenomenon at the crossing of these two and thus deepening understanding of the development of action competence and transformation of soldiership as an educational endeavor. In the following, our examination is located and anchored to the central concepts of military pedagogy, that is, action competence and cognitive readiness. After that we explore the experience of the unexpected and consider the possibilities and challenges of change in our habits of mind within military culture. We conclude the article by introducing the notion of "Revolution in Learning Affairs" which emphasizes an unutilized reservoir of human self-awareness to enhance military performance.

---

2   Mezirow's theory deals with the transformation of adults' meaning perspectives. In addition to examining this kind of transformation within a military context, we also read the theory backwards and extend it to a context beyond its normal range of application by examining the challenging issue of preparing for the unexpected on the basis of Mälkki (2010, 2011) and Mezirow (1991).

3   Cf. sports psychological considerations on the linkage between affect and performance as well as on managing the influences of stress to performance (Baumeister, 1986; Golden, Tenenbaum & Kamata, 2004; Forgas, 1995; Gould & Tuffy, 1996; Hanin & Stambulova, 2002; Johnson, Edmonds, Tenenbaum & Kamata, 2007; Jones, 2003).

Action competence and cognitive readiness – the beacons of transforming soldiership

On the individual level there are already beacons pointing to the direction and goals of such a transformation by highlighting the essential qualities of a modern soldier. Here we examine the issue with the help of the concepts of action competence by Toiskallio (2000, 2004, 2005) and cognitive readiness by Fletcher (2004). Action competence highlights a holistic conception of the human being as the basis for military pedagogy (Royl, 2005; Toiskallio, 2000, 2004, 2005). It includes practical wisdom and points to decision-making in situations where it is not enough to rely on application of formulas, but where instead we must be able to make decisions without complete information and act in unexpected situations. According to Toiskallio (2004), identity and self-knowledge are at the heart of action competence, and are the ultimate source of meaning and experience as they penetrate through all our knowledge. Although highlighting the soldier's competence from an individual viewpoint, Toiskallio (2000) also emphasizes that we are always within our social and cultural contexts that shape our identities and perception. However, he does not finish paving the road for us, but instead presents the following crucial questions (ibid., p. 58): "How is the action competence of soldiers (soldiers and officers at all levels) developed? And, what are the requirements of action competence in various practical tasks, and how are they reached?". As an answer this call, it is our purpose here to take part in the discussion by presenting a viewpoint concerning theorizing the development of action competence. We attempt to bring together the concepts introduced by Toiskallio, and to work out the relations between them in order to build a ladder to the transforming of soldiership and enhancement of the action competence of soldiers by focusing on the necessary understanding of what we are today, that is, our habits of mind that orient and limit our practical wisdom.

Another viewpoint concerning the competence requirements of a modern soldier is presented by J. Dexter Fletcher. His conception of cognitive readiness emphasizes the cognitive components that are essential for the soldier to be able to act creatively in complex, unexpected and chaotic situations. In these kinds of situations no previous training will fully suffice and one must respond immediately without being able to consult with senior officers. Fletcher's list of components of cognitive readiness includes situation awareness, memory, transfer, metacognition, automaticity, problem-solving, decision-making, mental flexibility and creativity, leadership including interpersonal competencies, and emotion (Fletcher, 2004). Cognitive readiness can be seen as a cognitive-scientific interpretation of action competence (Toiskallio, 2004). In addition, Fletcher (2004) acknowledges that we create the world through our perceptions and are thus reacting to a reality according to our construction of it. However, although Fletcher acknowledges this essential constructive character of our cognitive processing, he does not go further to examine the nature and orientation of this construction process that can be seen to affect all components of cognitive readiness.

Both of the above scholars highlight "preparing for the unexpected" as the guiding principle to be kept in mind in military education. However, this phrase is in many ways a slippery one. Once we have been able to achieve preparedness in terms of knowledge and expectations, things would no longer be unexpected. Actually, in order to structure our goals and methods we need to differentiate between the different uses of the phrase. Probably the more common way to interpret it is to highlight the difference between the unknown future and the presumably known present, and aim to bridge this gap of knowledge and understanding with education. Thus we would in a way try to predict the unexpected and diminish the chaos by controlling it (deductively), that is, we would try to make it more expected than unexpected.

The other way to understand 'preparing for the unexpected' would refer to 'preparing to *experience* the unexpected,' that is, a more inductive approach where the unexpectedness is accepted and we are ready to face the fact that our expectations will not be sufficient in order to understand what is happening. In this we are again at the heart of Lloyd's excerpt concerning whether we let our expectations dictate what we make of our experience/perception (deductively) or *whether we are able to use a more inductive view* to base our decisions on. The latter viewpoint at the same time opens a way to the basis of cognitive readiness by raising the question concerning the orientation of our cognitive capacities and knowledge construction. However, the purpose here is not to contest the idea of cognitive readiness but to approach it from another point of view and thus complement it and examine the basis of it. For Toiskallio (2000), the development of judgment is essential in enhancing action competence, as there is a challenge for the soldiers to be able to improve their abilities to interpret, handle and understand information. Thus a more detailed understanding of 'preparing for the unexpected' appears to be also at the heart of the development of action competence.

The human factor – experiencing the unexpected

Despite the fact that the unexpected has long been seen as a characteristic of war, the emotional-level meaning and experience of these kind of circumstances have rarely been paid attention to with regards to educational spheres. However, in order to understand the challenges of soldiers' action competence, we need to look more closely at the very notion of unexpectedness. How is it possible to know anything about the unexpected, or even to know about the experience of meeting with the unexpected? However, since we are talking about the unexpected, we are at the same time expecting it. Therefore it is not completely unexpected. What do we actually know about unexpectedness or chaos? Maybe it is easier to approach the issue from the opposite direction; that is to say, what is it when it is *not* unexpected or chaotic? Presumably it is expected and probably also very clear. Interestingly these words appear to come together in *understanding*: Expectedness can be experienced in a situation that is understandable, and on the other hand it can be regarded as unexpected or *chaotic* when we are not able to cope with an environment, when we do not understand what is happening either in a situation or within ourselves (Mälkki, 2010). Thus the situation is not *understandable* in terms of our previous experiences and *expectations* (Mezirow, 1991, 2000, 2009) and we may feel *anxious* and even fearful, as there is no sense of *safety* based on situations being *understandable* and the future being predictable in terms of our previous assumptions (Mälkki, 2010). More precisely, in the case of chaos we are not able to make meaning in the light of our *meaning perspective*, which is the orienting frame of reference or personal paradigm that comprises our values, attitudes, knowledge and feelings and is shaped by language, culture and personal experiences (Mezirow, 1991, 2000, 2009).

The basic logic of our mental functioning can be seen in this: we are able to understand things only within the light of our previous understanding; we in a way grasp the unexpected with our expectations, and the result is our subjective perception and interpretation of the situation (Damasio, 1999). This is also highlighted by Huhtinen (2004, p. 90-91), who mentions that "in the great blooming, buzzing confusion of the outer world we pick out what our culture has already defined for us, and we tend to perceive that which we have picked out in the form stereotyped for us by our culture". This refers to our basic predisposition to interpret in the light of our meaning perspectives. This oriented and subjective base for interpretation enables us to understand with the help of our previous experiences and our personal histories and expectations of the future. Further, it enables us to locate the experiences within our personal histories and experiences, since at the very core it yields us the awareness that it is I who is experiencing this. At the same time this subjective construction of meaning makes our view limited, simplistic and in a way biased. We block certain aspects out of our awareness in order to avoid anxiety and make situations seem more understandable. This orientation of our perceptions and interpretations results in a predisposition to think, feel and act within a *comfort zone*[4] of familiar meanings (Mälkki, 2010;

---

4 The term comfort zone is used for the level at which one functions with ease and familiarity, thus pointing to a more human scientific approach (Merriam-Webster, 2008). Within adult education, a

Mezirow, 1991, 2000, 2009). The frame of reference for understanding is our personal paradigm called the meaning perspective, and inability to understand or chaos is experienced as anxiety, i.e., edge-emotions[5]. Therefore, chaos can be seen to be more our subjective experience of *not being able to understand* than a characteristic of the external environment.

In addition to the formation of our perception and our understanding, the previous examination brings us to the *experience* of facing the unexpected, as emotions were shown to be deeply intertwined with the more cognitively focused interpreting and understanding. We may be unaware of our emotions when things are happening as expected, whereas emotions arise especially when our expectations do not yield us an understanding of the situation. Often we talk about the situations of war being unexpected and chaotic, which leads us to realize that we are forced to act in such situations along with knowledge gaps, and we may focus on our existing knowledge and the extent to where it reaches. This mostly epistemic and cognitive discussion is in need of a more experiential viewpoint in order to understand the case of making decisions in unexpected situations and to enhance the action competence of the soldier so that they may be able to cope with these kinds of situations (Toiskallio, 2004). How do we actually experience such situations? Are we comfortable because we *knew* that the situation was supposed to be unconceivable, or do we experience insecurity and a lack of competence when our expectations are recognized as insufficient in interpreting what is going on in the situation (Mälkki, 2012a, Tenenbaum et al., 2008)? And further, how do these feelings affect our decision-making?

---

mental safe territory, i.e., protective belt, is discussed in the work of Anita Malinen. She bases her view on an analysis of the adult experiential learning theories of Knowles, Kolb, Mezirow, Revans and Schön and concludes with the concept of personal experiential knowing, which orients one's thinking and thus 'saves the learner from becoming confused by the 'ocean of anomalies'." Consequently certain paths are avoided and the safe territory may be maintained by building up eclectic hypotheses. Within literature applying Mezirow's theory, the term comfort zone is generally used to refer to the limited set of behaviors in which one is comfortable (Malinen, 2000; Mezirow, 1991).

5 *Edge-emotions* is a concept introduced by Kaisu Mälkki (2010, 2011, 2012b; Mälkki & Mälkki, 2011) to represent the experiential or emotional dimension of the meaning perspective, in order to explicate the intertwinement of cognition and emotion in thinking and feeling. It involves explication on how the biological function of emotions in ensuring survival affect also our thinking in order to protect the intactness of our meaning perspectives. This concept was originally developed within research that aimed to understand and theorize the prerequisites and challenges to reflection in adult learning. It is theoretically rooted in Mezirow's theory of transformative learning and Damasio's neurobiological view on emotions and consciousness. For more detailed consideration, see Mälkki (2010, 2011).

A common way to view emotions is to regard them as basically negative and disturbing, something that must be controlled in order for one to be able to think rationally. As a consequence, emotions are viewed to be in need of controlling and excluding if soldiers are to perform complex tasks in stressful and confusing modern military contexts.[6] However, Antonio Damasio's recent brain research has shown that emotions have a fundamental and more complex meaning in respect to both our cognitive functions such as interpretation and decision-making, and to producing quick reactions in critical situations by directing attention. It is the emotions that direct our attention based on our previous experiences of on similar situations and at a very basic level support our life-support systems that aim at the maintaining of life (Berridge & Winkielman, 2003; Damasio, 1999). In terms of mental functions, emotions support the consistency of the meaning perspective by directing attention to the comfort zone and arousing negative feelings when the comfort zone is being exceeded. Thus the anxiety that arises when one is unable to understand, basically serves the consistency of consciousness by directing attention to the familiar aspects and by resisting questioning of our meaning perspectives that include the basic assumptions and values on which our identities are based (Mälkki, 2010, 2011; Mälkki & Mälkki, 2011). This brings us back to Lloyd's excerpt concerning the inductive interpretation of the situation. In the light of the discussion above it can be seen that the experience of unexpectedness is most of all an emotional experience of the threat of chaos resulting from being unable to understand. How are we actually going to cope with the unexpected if our understanding is leaning on the expected? According to Mezirow (2000, p. 3), "if we are unable to understand, we often turn to tradition, thoughtlessly seize explanations by authority figures, or resort to various psychological mechanisms, such as projection and rationalization, to create imaginary meanings". Using Lloyd's terms, this can be seen as a mechanistic application of rules and thus "making nature subservient" in order to create imaginary meanings and understandability which would make us feel more in safe (Malinen, 2000). The basic motivation for this simplistic interpretation and compulsive leaning on our previous expectations stems from our basic need to avoid anxiety and chaos, and to seek safety and the comfort zone (Greenberg & Paivio, 2003; Greenberg & Pascual-Leone, 2006; Mälkki, 2010, 2011). Thus our interpretation of the situation can be seen to serve more our need to bring about ostensible understandability and safety than our objective of inductive understanding of the situation

---

6 Cf. Fletcher (2004); within military psychology and sports psychology the complex role and inevitable presence of emotions has been acknowledged, and rather than emphasizing a mere control or aim to exclude emotions, attention is paid to awareness of the effects of these, aiming to utilize the emotions in performance by re-interpreting them (cf. Tenenbaum et al., 2008).

itself. The more emotionally threatening the situation is, the more likely we are to exploit reason in order to bring about safety through simplistic interpretation and thus might lose sight of the original task of understanding the environment.[7] Therefore when the comfort zone is being exceeded, the (edge-)emotions first and foremost motivate us to restore the balance, that is, to perceive and interpret things in the light of our expectations in order to *feel* the world as understandable. Thus the mere control of emotions would not help to allow inductivity in our thinking, or flexibility and ethicality in our practical wisdom. Rather it may decrease the flexibility even more if one's thinking is tied to controlling the unwanted emotions. In terms of the social dimension, there is similarly a great temptation for choosing collectively accepted alternatives that would not bring about social discrepancies and cause unpleasant feelings.

Preparing to exceed the comfort zone

The above-introduced nature of decision-making and interpretation in unexpected situations has profound implications for military pedagogy and the education of soldiers. It underlines the experience of facing the unexpected that can be seen to be at the heart of preparing for the unexpected. In order for the soldier to be able to act creatively, responsibly and ethically in unexpected situations, they must be able to cope with the edge-emotions, that is, become aware of and *understand the edge-emotions*, the anxiety and the feeling of being threatened by chaos that result from being unable to understand the situation. Anxiety and feeling threatened by chaos were shown previously to depend on our meaning perspective. Therefore our meaning perspective and the related edge-emotions appear as the very source to be worked on in order to prepare for the unexpected.

Working on our edge-emotions refers to accessing and engaging them and trying to understand them as being a result of our expectations appearing insufficient in explaining the situation. When the comfort zone is being exceeded, our thinking tends to narrow in favor of restoring the balance and returning to work within the comfort zone (Mälkki, 2010, 2011). However, the very emotions arising on the edges of the comfort zone are the key to both bringing understandability to a chaotic situation and being able to reflect on our assumptions in order to enhance

---

[7] For the effects of stress and affect on performance, cf. Tenenbaum et al (2008); also Bartone (2006); Baumeister (1986), Forgas (2005), Golden et al (2004), Gould and Tuffy (1996), Hanin and Stambulova (2002), Johnson et al. (2007), Jone (2003), Litz (2007) and Wallenius et al. (2004). In this article, however, it is focused on how affect may influence one's *interpretations* and how one may try to diminish the extent of these influences (cf. Mälkki & Mälkki, 2011).

our action competence and widen the scope of our cognitive readiness. In fact, the ingredients of the experience of chaos or anxiety can be seen to be the assumptions or expectations being challenged since they do not yield an understanding of the situation in the way one had expected. The edge-emotions in a way cover a contradiction or conflict within the meaning perspective. Thus becoming aware of edge-emotions and giving them meaning is necessary in order to reach the assumptions behind the emotions that limit our thinking.[8] Giving meaning to such emotions or trying to understand them paradoxically enables us to bring understandability to the chaos and thus makes it possible for us to diminish its effect and allow more capacity for decision-making and interpreting the situation.

However, as the basic function of these edge-emotions itself is to support the consistency of the meaning perspective by directing attention back to the comfort zone, they are not very easy to manage. As a matter of fact, we tend to direct attention away from them, create imaginary meanings in order to bring safety and understandability to a situation, or interpret the emotions as being an indication of the harmfulness of the situation.[9] Even though these predispositions have their base in the life-support system, only in our own culture we have *learned to neglect* these emotions and the crucial possibilities for learning and extending our comfort zone that they contain. This is precisely why we *can* also learn to accept and acknowledge them, and practice dealing with them in order to release our cognitive capacities in the cases of chaos.

---

8   Dewey and Mezirow, among others, consider problematic situations or disorienting dilemmas a trigger for reflection and transformation (Mälkki, 2012a; Toiskallio, 2000). Also, for Malinen a discrepancy, i.e., second-order experience, is essential for adult experiential learning, as it is seen to stimulate a shift in personal knowing. Second-order experience brings about doubt and negative feelings while realizing continuity in terms of the personal experiential knowing. However, in these views it is not acknowledged that the contradictions in assumptions are not automatically reachable in the supposedly rational analysis, but are intertwined with the emotional and social dimensions of meanings, which may also orient our thinking to avoid the contradiction itself. In order to understand the practical prerequisites and challenges as well as the obstacles of reflection, it is necessary to consider the emotional and social dimensions of reflection along with the cognitive. This redefinition is based on an elaboration by Kaisu Mälkki of Jack Mezirow's conceptions on reflection (Mälkki, 2010).

9   Cf. Mälkki (2010) and Damasio (1999). According to Damasio, we often interpret our emotions as being an indicator of the quality of the environment, for example whether we *think* the situation is good or bad for us or whether the decision being made seems to us good or bad. Many times this is very useful as our emotions automatically direct our attention and may cause us to want to avoid dangerous situations or to approach pleasant ones. However, this arousal of emotions is determined in relation to our meaning perspectives, and thus, instead of assessing the characteristics of environments, the emotions may also lead us to assess the situations or decisions based merely on our habits and personal preferences without being willing to consider other alternatives equally (Mälkki, 2010, 2011).

In fact, these emotional experiences at the edges of the comfort zone bring out a dimension of human experience that is similar in both war and peacetime contexts. This offers a link between wartime and peacetime contexts that are often seen as different spheres of a soldier's action. However, based on the above examination of military habits of mind, it is possible to discern situations with regard to whether the comfort zone is being exceeded and, on the other hand, whether one is physically safe or in danger. These two dimensions are linked together in the figure above. The vertical dimension represents physically experienced threat on the lower part and feeling of safety on the upper part. The contexts of war and peace may be seen as examples of these, respectively. However, in the war context it is also possible to feel relatively safe, if one is not in immediate contact with hostile intentions, for instance, and thus this would be placed on the upper part of the figure despite the context of war. There are fundamental differences between perception and understanding of warlike situations. That is to say, this dimension emphasizes the nature of the situation from the viewpoint of a certain individual rather than from a nation's, for instance.

Figure 1: The domains of military education and activity.

The source of experience of threat / safety

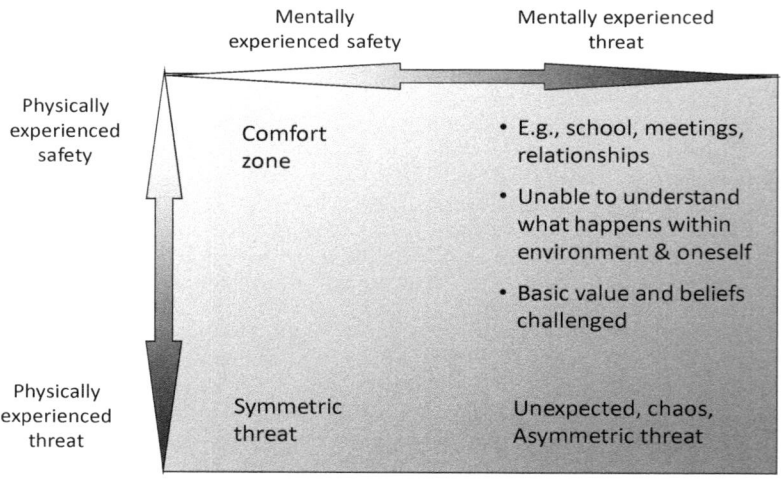

The horizontal dimension, on the other hand, represents the experience of the situation in terms of mental comfort. The right-hand side represents circumstances which we often consider "normal" in the sense that the situation is experienced to be expected enough and we are able to cope with it automatically and thus may not even be aware of the properties of the situation. That is to say, we are able to stay in our comfort zone as the world seems understandable and nothing challenges our basic beliefs and values. Respectively, the left-hand side represents circumstances in which the comfort zone, the experienced normal order of the world, is being questioned, for example since the situation is unexpected and we are not able to understand what is happening and to cope with it. A similar experience may also be caused if someone questions our basic assumptions and values.

By considering both of these dimensions, Figure 1 illustrates four different kinds of contexts and the similarities and differences between them. It is important to notice, however, that the classification draws exaggerated lines between differences among different contexts, for instance between set piece scenarios and unpredictable scenarios. However, this distinction is important in order to understand the difference between acting in expected and unexpected situations. Within set piece scenarios our life is threatened but we may, nevertheless, experience the situation as understandable. However, in the cases of unpredictable scenarios, in addition to the physical threat, we experience the situation as threatening also from the mental point of view, as we are not able to understand the situation and stay within the comfort zone of familiar meanings. As a consequence, unpleasant feelings, i.e. edge-emotions, are aroused, which orient us back towards the comfort zone. That is to say, we thus aim to restore the balance and understanding and aim to bring back the experience of being able to understand, as was shown earlier. This kind of situation brings about new challenges in terms of maintaining responsible and creative decision-making, as the circumstances orient us towards shortcuts to comfort instead of maintaining flexible thinking and rigorous judgment.

However, an interesting viewpoint to these chaotic situations inherent in battlefield conditions is opened by the similarity between the unpredictable scenarios of battle and the physically safe environments which nevertheless challenge our comfort zone (lower and upper left side, respectively). Despite the differences presented, these two contexts share crucial similarities that may be benefited from in military training. Both of these situations challenge the comfort zone and arouse edge-emotions, and consequently distract our judgment. Thus the very element that distinguishes unpredictable scenarios from set piece scenarios is nevertheless possible to experience in educational settings. Therefore it is possible to enhance one's practical wisdom and action competence of making decisions in the middle of unexpected situations by learning to work on edge-emotions in educational settings.

Biologically we have the emotional support for quick reactions in critical or dangerous situation, such as escape, attack, and searching for safety (Damasio,

1999). Thus, when soldiers act in situations that are physically dangerous, they are already facing unpleasant feelings that stem from our biological life-support systems aiming to warn us of danger. However, it is nevertheless a matter of our mental capabilities to make decisions and aim to act in terms of our objectives and strategies. In the previous paragraphs it was emphasized that our experience in the face of the unexpected is crucial in determining the flexibility of our thinking and decision-making. Therefore we need to be prepared to *experience* the unexpected in order to enable the best possible flexibility in our thinking. This is possible, for example, by learning to acknowledge the edge-emotions that are already present in our everyday actions. Another way is to challenge our normal ways of thinking that then trigger edge-emotions and enable us to learn to manage them. Either way, the idea in familiarizing ourselves with these emotions is that it may reveal the limits of our cognitive capability as well as the biases of our interpretation, as these emotions automatically orient our thinking towards the comfort zone and to neglect the complexity of the situation. Furthermore, these emotions are the gate-keepers of our long-held assumptions that orient our thinking in the first place. Thus this kind of training may also transform our meaning structures into more flexible ones, as we may learn to become more sensitive to the habits of mind that solidify our thinking when unquestioned.

Understanding the cultural level obstacles to adopting new ideas

Armed forces unavoidably follow the cultural habits and expectations of the surrounding society. It is extremely difficult for military organizations to adopt new methods concerning the art of war, i.e. the general idea of using military organization to gain advantage over the opponent. The art of war has its roots in the forms of cultural self-evidences, manners and habits (i.e., comfort zones) which produce concrete obstacles to adopting the essential core from the art of war. In particular, emotional-based learning methods may arouse subconscious or even uncovered resistance. One example is the German First World War military teaching and learning method *künstliche Aufregung* [artificial agitation], which was designed to artificially stimulate minds and bring about an atmosphere of chaos in the middle of military exercises. Finnish officers refused to put this method into practice domestically in the early 1920s because it did not suit the "national character" of the Finns (Malmberg, 1920). What was not understood was not carried through. Similar blocks and mental obstacles were created in front of any deep reforms concerning ideas on the art of war (Mälkki, 2008). Leadership methods and practices hidden in the system of discipline are especially more difficult to alter. Problems arise particularly when organizations try to locate their habits of mind, as it is extremely difficult to see the actual appearance of them through the lens of culturally constructed meaning-making. The attempts are even more difficult if the objective is to understand adversaries and their military thinking, or to locate the nature of their art of war, the practice they are following.

Naturally, any standing army needs collective and officially approved habits of mind and mindset, but also a clear mutual understanding of how and why things are done in a particular way. This is extremely important because military effectiveness is based on ensured collective behavior, formalities and traditions. However, military concepts are not just clichés to military organization. It is preparing to face the utmost extremes in the battlefield while it must be ready to subdue the enemy's will to resist or even to destroy the enemy. The core of military know-how is based on how certain commonly and widely used military concepts, i.e. military terminology, are understood and also emotionally felt. For professional soldiers, words like "defense," "attack" or "ambush" include certain emotional presumptions, as they involve certain "feelings." Therefore any major organizational "transformation" is likely to face resistance – at least, if the modification is attached to patterns of behavior or widely used military terms or concepts. Resistance is not just an act of insubordination. It is rather an attempt to keep the basic military structure and its "way of doing things" untouched as well as an attempt to keep concepts comprehensible enough (Mälkki, 2008). At the final stage of the Second World War, American observers did not understand the importance of German stubbornness in defending some islands in the English Channel: The observers could find no rational reasons for that kind of behavior, as this defense lacked sensibility in military strategic point of view. Very likely the Germans were practicing distributed command (*Führen mit Auftrag*), which was unfamiliar to the Americans at that time. German military education had produced a functionality based on meaning-making that was absorbed mentally (emotionally) as well as physically. The commanders had absorbed the idea of "defense" in a way that they (probably) could have felt physically, especially if there was a danger that the mission might lead to failure. Merely the fear of this could launch an emotional reaction of "disgrace", which probably improved battle performance. Cultural ignorance was reciprocally felt. German observers (POWs) did not understand how the American military system could operate at all (Shils & Janowitz, 1948).

Transformation of soldiership

The cultural level barriers to adopting new ideas discussed above at the same time call for understanding of the individual challenges of transformation. In terms of action competence and coping with complex situations, two essential questions need to be addressed: First, what is the width, breadth and flexibility of our meaning perspective that defines our comfort zone, the scope of our cognitive functions as well as the boundaries of our practical wisdom? Second, *how do we cope with the anxiety* that arises when we do not understand or when things do not happen as expected, that is, when the comfort zone is being exceeded? In a more context-specific way we need to ask: What does the culturally and organizationally shaped meaning perspective of a soldier look like? What are the limits of the military mind? How do the military organization, culture and art of war shape the meaning perspective of a soldier? And further, what kind of influence does this have on action competence and performance in unexpected situations?

In order to learn to manage something that is on the edges of our comfort zone, we need transformative learning in the literal sense of the word. Developing action competence in these terms cannot be a matter of the more common (assimilative) way to learn, i.e., adding elements into and according to our prevailing meaning perspectives and habits of mind. To the contrary, we need to be able to *transform* the very meaning perspective that orients our normal thinking and keeps us within the self-evident normal conceptions of normality that limit our scope of cognitive functions, thinking, feeling and acting (Florian, 2000; Illeris, 2004, 2007; Kegan, 2009; Mälkki & Mälkki, 2011; Mezirow, 1991), that is, our lines of natural expectation. Our habits of mind, conceptions of normality and most of all our emotional patterns are based on the cultural self-evidences that also mirror the values and habits of the previous generations. When we are socialized into our culture, we absorb and acquire the emotional patterns and values even though on the conscious level we are living in a different time and assume that we are different from the generations before us (Brookfield, 2000; Mälkki, 2008; Mezirow, 1991;). However, in terms of learning and development these dispositions make us more likely to learn in order to confirm our current meaning perspective rather than to aim to transform it (Mälkki, 2010, 2011). How is it then possible to bring about transformation, if we have a predisposition to protect our existing habits of mind and resist challenging them? Transformation presupposes becoming aware of our current meaning perspective and questioning it, in other words, critical reflection (Mezirow, 1991, 2000, 2009). However, the intactness of the meaning perspective is protected by the anxiety (i.e., edge-emotions) that arises at the edges of comfort zone, thus resisting critical questioning of assumptions. Therefore the key to transformation can be seen to be the very same edge-emotions that previously were shown to be at the core of preparing to experience the unexpected.

Thus, in terms of preparing for the unexpected and enhancing action competence, we need a transformation of soldiership on two intertwined levels. First, we need to be able to transform the very habits of mind and the emotionally anchored conceptions of normality that orient us towards the past wars and thus prevent us from being sensitive to future unexpectedness. Second, on the emotional level we need to learn and practice to becoming aware of and managing the painful feelings (i.e., edge-emotions) that precisely aim to preserve our prevailing habits of mind by directing attention back to the comfort zone of normal thinking. Thus the two levels of transformation come together in the edge-emotions which are the very elements that guard our prevailing habits of mind and prevent us from extending our view.

The ethical core of soldier's action competence

The above levels defining the transformation of soldiership are also closely related to the objective of the soldier being an ethical and responsible agent. The grounds for our decisions are hidden from ourselves if we do not understand the influences of the emotions. That is to say, how can we be responsible agents and decision-makers if we are not aware of the cognitive and emotional bases that direct our perceptions and interpretations (Pulkka & Raviv, 2007; Toiskallio, 2005)? We have an amazing capacity or potential to be able to think about the way we feel, and this should be enhanced and practiced within military education. Soldiers' spheres of activities are the extreme contexts in battle areas where normality and comfort zones are often exceeded. These instances are always emotional matters, as our basic mental functions work according to our habits of expectation and emotional patterns, and these have their own intrinsic agendas of maintaining life and ensuring comfortable mental functioning by supporting the meaning perspective and staying in the comfort zone.

We do not understand the unexpected by virtue of our normal thinking that aims to keep us within the comfort zone. Aiming to understand the unexpected requires sensitivity to perceiving something that does not automatically yield meaning to us. We must be able to cope with our personal insecurity because we experience a lack of competence when our expectations are recognized as insufficient. In order to cope with the unfamiliar we must start by confronting the *unfamiliar within ourselves*, that is, what to us seems self-evident to the extent that we are not even aware of it – our most natural ways of thinking, feeling, interpreting and acting that we have acquired in socialization and through personal experiences (Florian, 2000). What are the ways of working that to us seem to need no grounds but appear as justified by virtue of being generally accepted as well as common sense? What are the building blocks of our identities, the building blocks of our organizational values, the organizational cornerstones that within us take the form of emotional responses in order to favor certain viewpoints and to avoid and deny others?

This also brings out the ethical responsibilities of the educator. Working on the edges of the comfort zone is a possibility for learning that, however, can be easily taken advantage of and exploited in terms of manipulation and purposefully causing pain to the learners. Therefore the implied values of personal growth, ethicality and responsibility apply most of all to the educator (Mezirow, 1991). Becoming aware of one's feelings and emotions as well as the assumptions behind them is the starting point and the objective of both the educator and the learner, but no norms can be stated concerning the kind of thoughts and feelings this kind of learning may provoke. Thus those within the hypothetically manipulative educator's range can also figure out the manipulative logic of the teacher and bring that up along with other experiences.

Revolution in learning affairs?

It is possible to detect two lines of training that are crucial for any military. It is also important to distinguish between them (Tenenbaum et al.,2008). The first line of training concerns ensuring functional action by familiarizing the soldier with the strenuous circumstances and patterns of action that also bring about coherence in the midst of chaos. The second line of training, on the contrary, aims to prepare the soldiers to be able to act in situations when these patterns and habitual actions are not enough to be able to cope with the situation and to decide on the course of action, but judgment and contextual understanding is needed.

The first line is related to the fact that all regular armies are training their soldiers to face the chaotic battlefield conditions somehow. Usually particular and appropriate functions, action or motions are repeated hundreds of times in order to achieve the level of muscular memory. These functions must be trained constantly if they are to be performed in battlefield conditions where intensive and even paralyzing stress level may block all cognitive capability and even normal bodily functions. Stress level may raise the beat of the heart as high as 180 per minute, even without any physical activity (Grossman, 2007; Tenenbaum et al., 2008). These kinds of bodily experienced conditions may be seen as direct consequences of the real battle.

The second line of military training complements the first line by focusing on the mental flexibility of the soldier to be able to act in situations that may not be managed with previously rehearsed patterns of action. Broadening one's comfort zone and learning to cope at the edges of it is important for any professional. However, for a soldier facing the extreme conditions of war, this may be seen to be of even greater significance. In fact, the conditions of war may be seen to challenge one's comfort zone in unforeseeable ways and so it is a matter of training to diminish the restrictive effects that these harsh conditions bring about to our judgment, understanding and decision-making. In addition, it must be noted that the organizational culture and the accompanying patterns of behavior play also a role in determining the ways in which we habitually manage these situations.

Soldiers, and especially professional ones, need continuous mental, emotional, intellectual and spiritual training to be able to make sense of what they do not at first understand, or what they might bypass over subconsciously. Every method which supports soldiers in practicing self-directed discipline helps. For example, simulators may be used as stimulators to help individual soldiers, teams or groups find their self-evident assumptions and their comfort zones. This state of affairs may cause unpleasant feelings, especially when situated in the middle of a familiar, routine-based exercise. Emotions rise to the surface if the simulated exercises do not lean on meanings which are familiar to the soldiers. The most effective way to use simulators is *not* to use them to train to kill or destroy more efficiently, but to learn to handle uncomfortable situations which could lead to unconsidered behavior and decisions which are not ethically justified. We need to work with our cultural and mental Achilles heels to be able to handle and control our inner-directed behavior which may be more familiar to our adversaries than to us. This kind of training also supports the qualities that are essential for recovering from post-traumatic stress disorder (Grossman, 2007; Palosaari, 2007). In therapeutic treatment the ability to acquire a depth of emotional processing and to increase it, together with reflection on emotions, has been shown to be related to good therapeutic outcomes. Attending to one's emotions and conceptualizing them enables one to reflect and to create new meanings to explain the experience as well as to share one's experience with others (Greenberg & Paivio, 2003; Greenberg & Pascual-Leone, 2006). Learning to make meaning and to process one's emotions can thus be important for the soldier in order to understand and be able to process his or her experiences alone or with others both while on duty and after home-coming.

Naturally, military effectiveness is based on ensured and especially sufficiently predictable collective courses of action. This is the lifeblood of a military organization, as it must create certain habits of expectations, routines and collectively accepted "ways of doing things", in order to maintain its functionality during extreme battlefield conditions. Waging war is a serious and unpredictable business and therefore simple but functional forms of behavior, action, regular patterns, as well as collective and officially approved habits of mind and mindset are of great value. Clear methods of leadership and control methods were needed in battlefield conditions in order to handle chaotic situations. During the 19th century there appeared to be no need to instruct leaders on the unpredictability of the battlefield because they learned to handle unfamiliar and chaotic situations through experiencing the real battle. Training methods were therefore focused on ensuring that troops were disciplined enough, as there was seemingly no need to train for tolerating uncertainties and the emotional and cognitive reactions brought about by the ultimate experiences of battle. This kind of attitude to instruction remained intact through the centuries. We are still, in the middle of our technological innovations, almost incapable of reaching the essential core of military education. That is to say, we are lacking sufficient methods to be able to reach the circumstances similar to mental and physical threats.

We are definitely focusing too much on making our future soldiers interoperable land warriors equipped with data- and sensor fusions. Besides developing the technical capability of soldiers, we should focus all the more on instructing soldiers to be more ethically competent. The true challenge for any professional soldier is to achieve sufficient level of knowledge on the subject. On the other hand, we may question whether the professional soldiers are more competent in the field of ethical activities if compared to normal citizen-soldiers? What are the true boundary lines of civil and military education and the methods of teaching the subject?

Transformation is needed, as traditional European soldiership is still emotionally closer to "trench mentality" than to the abilities needed in dispersed operations (DO). It is not that we do not want to change – rather we may have insufficient means to make real changes happen. Modernizing the current methods of learning and teaching is perhaps not enough – perhaps we need a Revolution in Learning Affairs (RLA). Military cultures may not contain the easiest foundations for any kind of mental transformation as collective habits of minds and mind sets are bound in the traditions and in rigid disciplinary methods. During the Napoleonic era, there was only minor importance on the education of individual combatants who were seen to be group players in large formations. These formations were constructed from living bodies and the technology played important but not essential part of the actual fighting capability. Nevertheless, there has been no real change, transformations not to mention revolutions, concerning *how the actual fighting influences the living human body*. It is highly probable that even the future soldier will experience this dramatic occurrence similarly, especially, if they are still defined as human beings and not as machines.

Bibliography

Bartone, P. (2006). Resilience Under Military Operational Stress: Can Leaders Influence Hardiness? *Military Psychology*, 18 (suppl.), 131-148.

Baumeister, R. & Showers, C. (1986). A review of paradoxical performance effects: Choking under pressure in sports and mental tests. *European Journal of Social Psychology*, 16, 361-383.

Berridge, K. & Winkielman, P. (2003). What is an unconscious emotion? The case for unconscious "liking". *Cognition and Emotion*, 17, 181-211.

Brookfield, S. (2000). Transformative Learning as Ideology Critique. In J. Mezirow and Associates (Eds.), *Learning as transformation. Critical perspectives on a theory in progress* (pp. 125-148). San Francisco: Jossey-Bass.

Carrick, D., Connelly, J., & Robinson, P. (2009). *Ethics Education for Irregular Warfare*. Farnham: Ashgate.

Damasio, A. (1999) *The feeling of what happens. The Body and Emotion in the Making of Consciousness.* New York: Hart Court Brace.

Fletcher, J. (2004). Cognitive Readiness: Preparing for the Unexpected. In J. Toiskallio (Ed.), *Identity, Ethics, and Soldiership* (pp. 131-142). ACIE Publications, No 1. Helsinki: National Defence College.

Florian, H. (2000). Military pedagogy and *Bildung*. An Interview with Colonel Heinz Florian. (Interview by Jarmo Toiskallio). In J. Toiskallio (Ed.), *Mapping Military Pedagogy in Europe. Series 2, No 7* (pp. 11-18). Helsinki: National Defence College.

Forgas, J. (1995). Mood and judgment: The affect infusion model (AIM). *Psychological Bulletin, 17*, 39-66.

Forstén, O. (2005). *SE Analysis and After Action Review*. Tuusula: Finnish Defence Forces, Education Development Centre.

Golden, A., Tenenbaum, G., & Kamata, A. (2004). Performance zones: Affect-related performance zones: An idiographic method linking affect to performance. *International Journal of Sport & Exercise Psychology, 2*, 24-42.

Gould, D. & Tuffy, S. (1996). Zones of optimal functioning research: A review and critique. *Anxiety, Stress and Coping, 9*, 53-63.

Greenberg, L. & Paivio, S. (2003). *Working with emotions in psychotherapy*. New York: Guilford Press.

Greenberg, L. & Pascual-Leone, A. (2006). Emotion in Psychotherapy: A Practice-Friendly Research Review. *Journal of Clinical Psychology: In Session, 62*(5), 611-630.

Grossman, D. (2007). *On Combat. The Psychology and Physiology of Deadly Conflict in War and in Peace.* Milstedt: Warrior Science Publications.

Hanin, Y. & Stambulova, N. (2002). Metaphoric description of performance states: An application of the IZOF model. *The Sport Psychologist, 18*, 396-415.

Hartman, U. (2000). Military Pedagogy in Germany. In J. Toiskallio (Ed.), *Mapping Military Pedagogy in Europe. Series 2, No 7* (pp. 23-32). Helsinki: National Defence College.

Huhtinen, A.-M. (2004). Soldiership without Existence – The Changing Socio-Psychological Culture and Environment of Military Decision-Makers. In J. Toiskallio (Ed.), *Identity, Ethics, and Soldiership* (pp. 75-106). Helsinki: National Defence College.

Illeris, K. (2004). Transformative Learning in the Perspective of a Comprehensive Learning Theory. *Journal of Transformative Education, 2*(2), 79-89.

Illeris, K. (2007). *How We Learn. Learning and non-learning in school and beyond.* London: Routledge.

Illeris, K. (Ed.) (2009). *Contemporary theories of learning. Learning theorists...in their own words.* London: Routledge.

Johnson, M., Edmonds, W.A., Tenenbaum, G., & Kamata, A. (2007). The relationship between affects and performance in competitive intercollegiate tennis: A dynamic conceptualization and application. *Journal of Clinical Sports Psychology, 1,* 130-146.

Kegan, R. (2009). What "form" transforms? A constructive-developmental approach to transformative learning. In K. Illeris (Ed.), *Contemporary theories of learning. Learning theorists...in their own words* (pp. 35-52). London: Routledge.

Kiras, D. (2008). Irregular Warfare. In D. Jordan et al. (Eds.), *Understanding Modern Warfare* (pp. 224-291). Cambridge: Cambridge University Press.

Jones, M. (2003). Controlling emotions in sport. *The Sport Psychologist, 17*, 471-486.

Liddell Hart, B. (1991). *Strategy. Second revised edition.* London: Faber & Faber.

Litz, B. (2007). Research on the Impact of Military Trauma: Current Status and Future Directions. *Military Psychology, 19*(3), 217-238.

Lloyd, H. (1766). *The History of the Late War in Germany between the King of Prussia, and the Empress of Germany and Her Allies.* London.

Lonsdale, D. (2008). Strategy. In D. Jordan et al. (Eds.), *Understanding Modern Warfare* (pp. 14-63). Cambridge: Cambridge University Press.

Malinen, A. (2000). *Towards the essence of adult experiential learning. A reading of the theories of Knowles, Kolb, Mezirow, Revans and Schön.* Jyväskylä: Sophi, University of Jyväskylä.

Malmberg, L. (1920, January 5). *Selonteko suomalaisten upseerien opintomatkasta Englantiin.* Kansio 5, (Finnish) National Archives.

Mälkki, J. (2008). *Herrat, jätkät ja sotataito. Kansalaissotilas- ja ammattisotilasarmeijan muodostuminen talvisodan ihmeeksi 1920- ja 1930 -luvulla* [Gentlemen, Lads and the Art of War. The Construction of Citizen Soldier- and Professional Soldier Armies into "the Miracle of the Winter War" during the 1920s and 1930s]. Doctoral dissertation (monograph), University of Helsinki, Faculty of Social Sciences, Department of Social Science History.

Mälkki, K. (2010). Building on Mezirow's Theory of Transformative Learning: Theorizing the Challenges to Reflection. *Journal of Transformative Education, 1,* 42-63.

Mälkki, K. (2011). *Theorizing the Nature of Reflection.* Doctoral dissertation, University of Helsinki, Institute of Behavioural Sciences, Studies in Educational Sciences.

Mälkki, K. (2012a). Rethinking Disorienting Dilemmas within Real-Life Crises: The Role of Reflection in Negotiating Emotionally Chaotic Experiences. *Adult Education Quarterly, 62*(3), 207-229.

Mälkki, K. (2012b). What Does it Take to Reflect? Mezirow's Theory of Transformative Learning Revisited. *Lifelong Learning in Europe, 17*(1), 44-53.

Mälkki, K. & Mälkki, J. (2011). The Dynamics of Clausewitzian Friction. *Kungliga Krigsvetenskapsakademiens Handlingar och Tidskrift, 215*(2), 41-60.

Mäkinen, J. (2006). The Learning and Knowledge Creating School. Case of the Finnish National Defence College. *Tiede ja ase, 64,* 149-159.

Mezirow, J. (1991). *Transformative Dimensions of Adult Learning.* San Francisco; Oxford: Jossey-Bass.

Mezirow, J. (2000). Learning to think like an adult. Core concepts of transformation theory. In J. Mezirow and Associates (Eds.), *Learning as transformation. Critical perspectives on a theory in progress* (pp. 3-33). San Francisco: Jossey-Bass.

Mezirow, J. (2009). An overview on transformative learning. In K. Illeris (Ed.) *Contemporary theories of learning. Learning theorists ... in their own words* (pp. 90-105). London: Routledge.

Palosaari, E. (2007). *Lupa särkyä. Kriisistä elämään.* Helsinki: Edita.

Paret, P. (1990). Napoleon and the Revolution in War. In P. Paret (Ed.), *Makers of Modern Strategy. From Machiavelli to the Nuclear Age* (pp. 123-142). New York: Oxford University Press.

Pulkka, A-T. & Raviv, A. (2007). Military Ethics Instruction – The Educational Challenge of the Case-Study Method. *Tiede ja ase, 65,* 345-362.

Royl, W. (2005). Military Acting in the Spirit of Moral Obligations. In E. Micewski & D. Pfarr (Eds.), *Civil-Military Aspects of Military Ethics (Vol. 2). (Military) Leadership and Responsibility in the Postmodern Age* (pp. 63-73). Vienna: Publication Series of the National Defence Academy.

Shalit, B. (1988). *The Psychology of Conflict and Combat*. New York: One Madison Avenue.

Shils, E. & Janowitz, M. (1948). Cohesion and Disintegration in the Wehrmacht in World War II. *Public Opinion Quarterly, 12*, 280-315.

Toiskallio, J. (2000). Military Pedagogy as a Practical Human Science. In J. Toiskallio (Ed.), *Mapping Military Pedagogy in Europe* (pp. 45-64). Helsinki: Finnish National Defence College.

Toiskallio, J. (2004). Action Competence Approach to the Transforming Soldiership. In J. Toiskallio (Ed.), *Identity, Ethics, and Soldiership* (pp. 107-130). Helsinki: National Defence College.

Toiskallio, J. (2005). Ethics, Military Pedagogy, and Action Competence. In E. Micewski & D. Pfarr (Eds.), *Civil-Military Aspects of Military Ethics (Vol. 2). (Military) Leadership and Responsibility in the Postmodern Age* (pp. 132-143). Vienna: Publication Series of the National Defence Academy.

von Clausewitz, K. (1989). *On War*. Edited and translated by Michael Howard and Peter Paret. New Jersey: Princeton University Press.

Wallenius, C., Larsson, G., & Johansson, C. (2004). Military observers' reactions and performance when facing danger. *Military Psychology, 16*, 211-229.

Hubert Annen

Managing the Grey Area – The Meaning of Rituals in a Military Context and Their Effects on the Cadre's Responsibility

Introduction

A part of military training and education deals with learning what is right and wrong. This makes sense since a clear picture of the enemy or the task is necessary in order to take quick action or the control of a situation. The reduction of complexity creates a safe feeling of doing the right thing in an often ambiguous environment. Therefore one might sum up that on a basic level, military culture tends to foster black-and-white distinctions. What is prerequisite to be successful in military missions in many cases has also its flipside, though. This kind of perception leads to an unrealistic clarity, i.e., an all too easy picture of mechanisms in human communities, which in turn might engender inappropriate reactions.

With regard to their subordinates military leaders sometimes have the perception that such clear-cut distinctions exist in reality. There are at least two reasons for this. First, they generally have limited contact with the day-to-day realities of lower subordinate levels in their organizations and, as a result, have become unaware of what actually happens there. Second, reports superiors receive are usually filtered. Critical points have been eliminated and aspects which match their expectations have been emphasized.

From a (military) pedagogical standpoint it is dangerous to ignore the complex realities of human behavior. Such situations are seldom black-and-white. In every day (military) life there is always a grey area where decisions and actions depend on the particular situation and the people involved.

On the basis of such insights, this article deals with an aspect of human behavior, which is a characteristic part of military culture, i.e., rituals. Anyone who is genuinely familiar with military organizations should appreciate the importance of rituals. They help build cohesion and create an 'esprit de corps'. They have motivational effects. They can be crucial to enabling soldiers to confront dangerous situations. Unfortunately, however, they can also lead to actions that compromise people's dignity and integrity.

On Sense and Nonsense of Rituals

Recent media reports of ethically questionable initiation ceremonies in military training programs have again called public attention to rituals in the military context (Studer, 2008). The negative attention generated by sensationalist reports is likely to tempt military commanders either to ban such ceremonies outright or to deny their existence, dismissing the reports as the work of anti-military ideologues. However, rituals are part of everyday military life. Indeed, they have always been integral to human society. Therefore, it is necessary to examine rituals in a more differentiated manner so as to benefit from their positive aspects and to control their negative ones.

Accordingly, after examining the phenomenon of military rituals by analyzing the general role of rituals in human society, this article discusses the significance of rituals in a military context from a predominantly psychological perspective. It concludes by offering suggestions for addressing rituals in military education. First, however, it is necessary to define the character of a military ritual in our sense.

Definition of Ritual

The word ritual has many meanings and usages. Oftentimes, even simple repetitive activities such as salutation forms or the after-meal coffee are called rituals. The *Oxford English Dictionary Online* provides definitions that relate to actions prescribed by "social convention or habit", to practices of "religious or other devotional [or] solemn service", to "ceremonial acts", to "actions compulsively performed" (a common usage in psychology), and to actions "of a formulaic or repetitive type" (2008). Here, the meaning of ritual is restricted to actions that include a *metaphysical component*. This is the critical distinction between rituals in the sense of this discussion and other kinds of rituals. This metaphysical component is especially evident in religious observances. While military rituals may lack religious content, they do have a metaphysical essence, which can be found in the form of a common belief in something abstract and superior.

The nation, freedom, democracy – these are the types of ideas to which allegiance is pledged and for which one is willing to stand in formation or to march for hours (Euskirchen, 2005a). It is within this psychological framework that the *demonstrative* character of military rituals is evidenced. At the same time, military rituals have an introversive *forming* character. That is, they may aim at disciplining and preparing the individual in order to focus his strengths on a task that might require his absolute obedience, including suppressing concern for his personal well-being.

Rituals may also have a hybrid character, combining both demonstration and forming. These include rituals which confirm the individual's incorporation into the group. They serve to demonstrate to others that the individual is a member of the group, while, at the same time, help form in the individual the belief that he is a member of the group. Rituals which confirm group membership are called *rites of passage*. They seem to respond to a supremely human need. Therefore, they shall be addressed in detail.

Rites of Passage – A Cultural Anthropological Explanation

Human life progresses through stages of development. Societies often use rituals, rites of passage, to highlight the transition from one stage to the next. Such rituals emphasize the transition's significance and imbue in the individual a powerful memory of the transition. Particularly common among various societies are rites of passage acknowledging birth, sexual maturity, reproduction, and death. The significance of such "life crises" varies from one culture to another. In some cultures, a given transition may have little or no significance, while in others it may be celebrated with great pomp, including sumptuous rituals. The common thread is that, to our knowledge, no society ignores all such transitions.

Rites of passage were first systematically analyzed and scientifically classified early in the 20th century. It was recognized that such rituals help to overcome *critical periods* in human life (Bock, 1969). These critical periods encompass three stages: *separation*, *transition*, and *incorporation*. In the first stage, separation, the individual abandons his or her former social identity or status. That is, the individual is disassociated from his or her social position, either physically or symbolically (Bock, 1969).

The second stage, transition, is a kind of initiation. In that stage the individual becomes an *initiate,* i.e., the person to undergo the initiation rite, and is temporarily placed in exceptional circumstances where rules are suspended that are otherwise acknowledged. These circumstances constitute a psychological and social borderline state which has been described as one in which the individual floats between two worlds, the old and the new (Van Gennep, 1909; 1986). The initiates become members of their own social group, a so-called *communitas*, i.e., a tight-knit and exclusive community (Turner, 1969) existing at the edge of the greater social order. During initiation, the initiates may be subjected to extreme physical and/or psychological experiences, which depersonalize and instill fear in them. They are expected to draw strength from their ties to the collective and to the *communitas* to help them overcome their fears and depersonalization.

In the third stage, incorporation, the individual is freed from the exceptional circumstances of initiation and is accepted by the collective in his or her new social position, one defined by the collective's social system. Initiates enter this last phase only after they have received new rights and duties, i.e., after being *rehabilitated*. Accordingly, rites of passage usually lead to a fundamental change in the individual's self-conception.

Their shared borderline experience binds the group of former initiates together even after their re-incorporation into the society as a whole. They will remember the unusual, even uncanny, experiences they have undergone together for the rest of their lives. Thus, initiation produces social cohesion by inducing *amazement* and a sense of *communitas* in the initiates (Turner, 1964). This is why some researchers emphasize the significance of the *"community-endowing experience of collective distress"* (Streck, 2000).

Rites of passage are by no means relics of underdeveloped cultures and population groups (Schmid & Kocher-Schmid, 1992). They can be found, for instance, in school graduation observances from kindergarten through university level, at promotions, in citizenship ceremonies and pledges of allegiance, or on entry into the armed forces, the police corps, or the acceptance of any other public charge. They can also be found in the sometimes perilous self-initiation rituals of adolescents in the modern urban world, such as tests of courage, drug use, and gang warfare. What they all have in common is the educational function of initiation. It consists of *learning a new role*, and it certifies the *status of the individual* within a group.

The Function of Rituals in the Military Environment

This integrating function is of the utmost importance in the military context, and therefore it is the subject of regular military-critical discussions, whereby terms such as "subservience techniques" and "obedience production" are used (see Euskirchen, 2005b). Ultimately, however, it is the duty of any armed force to incapacitate a possible enemy by virtue of efficient exertion of physical force, or at least to emanate a credible determent. Hence, even in an era of increasingly engineered warfare, the action of a soldier remains the battle of life and death or simply the preparation for it (see Bröckling, 1997). With respect to such extreme experiences, a *special, adequate socialization* of military personnel is arguably indispensable. The issue here is to ensure the external control on the soldier on the one hand, i.e., he has to be trained to carry out orders from his superior, completely, conscientiously, timely, and with due commitment (Schweizer Armee, 2004). On the other hand, this requires the development of a rigorous and all-embracing self-control of the soldier, as he needs to meet the requirements imposed by his mission at any rate, against his very own fears and convictions, even when facing death.

Exercise, enforcement and training of certain values and (group-)standards, as well as systematic application of stress, are means of exerting influence on both attitude and tenor of military personnel. It is not uncommon that such educational measures and trainings are placed within the framework of a ritual, or that they adopt a ritual-like form themselves. Examples are ceremonies in connection with the inauguration of new combat units or with the incorporation and/or promotion of members, the almost metaphysical character and meaning of endurance tests in cadre trainings, or simple marches and assembly exercises at the beginning and end of a manoeuvre. On a factual basis, such actions grant *security* and *confidence* in one's own abilities and full functionality of the group. With regard to a combat situation, where success is by no means assured and which may even take a fatal course, emotional and ritual aspects tend to come to the fore. Simultaneously, respective activities take on irrational forms. Uncertainty, inexplicable and contradictory issues are channeled onto a metaphysical level in order to cope with fears and doubts and to tie together all participants by means of such a celebration.

What may look slightly abstract here can be suitably illustrated by drawing an analogy between military groups and a sports team in a similar situation, e.g., before an important game. Despite meticulous preparation in all areas, the members of the team cannot be sure about their success. Moreover, in order to win, they frequently have to take risks regarding their health. It is therefore of little surprise that in times of utmost uncertainties, repetitive rituals can be observed, sometimes almost reminding of magical thinking. Their purpose is obviously the strengthening of confidence in one's own capabilities and in those of the entire team, as well as the mitigation of mental stress to a bearable level (Baumann, 1993).

It seems therefore clear that rituals have a stabilizing function with regard to the mission, even if they appear irrational. The counter-rational component, however, holds the danger of developing their own *internal dynamics* since it works even without direct relation to the originally relevant circumstance. Consequently, it risks becoming a mere end in itself. Moreover, soldiers not only obey standing orders and orders of their superiors. Social control and affective conditions among comrades are equally important and decisive for both the cohesion of the troops and their disposition to violence. Initiation rites, informal codes of honor and sanction mechanisms, as well as informal, excessive alcohol consumption or even rape produce a feeling of conspired community (Winslow, 1999). *Conformity pressure* among comrades can accomplish what the authority of superiors alone cannot, since the individual member relies existentially on the support and goodwill of the group, especially in combat situations. Against the background of such insight, superiors have an obligation and responsibility to keep group dynamics under control, in order to prevent rituals from becoming ends in themselves but rather to make sure they have a *constructive* impact on both the mission and the people involved.

Consequences for Military Education

No one wants to experience within his field of responsibility rituals for the coercive "integration" of outsiders which comprise degrading means, sexual imagination, ceremonies that include elements reminiscent of black masses, or superiors who compensate for their personal shortcomings with ritualistic displays of power. Likewise, however, it is a fact that no armed force in the world can do without any rituals. Hence, neither negation nor suppression is an option; people in leadership positions should rather tackle these issues actively and constructively. To prevent the threat of rituals developing their own, independent dynamics, superiors should ensure that symbolic acts remain meaningful and *valuable* elements of military instruction, and particularly that such events maintain a worthy ritual framework. Statements such as "we always did it this way in our unit" and borrowed ideals from war movies have to be reflected critically. Ultimately, the goal is to understand the relationship between rituals and reality, mission, requirements of standing orders, and values of military education at all times. In this connection, it may even be advisable to include subordinates in the development and embodiment of respective rituals. This approach would foster the emergence of solidarity communities and helps achieve the target attitude of collective strength.

Bibliography

Annen, H., Steiger, R. & Zwygart, U. (2004). *Gemeinsam zum Ziel. Anregungen für Führungskräfte einer modernen Armee.* Frauenfeld; Stuttgart; Wien: Huber.

Annen. H. & Jufer, H. (2005). Vom Sinn und Unsinn von Ritualen. *Allgemeine Schweizerische Militärzeitschrift, 171* (10), 8-9.

Baumann, S. (1993). *Sportpsychologie.* Aachen: Meyer & Meyer.

Bock, P.K. (1969). *Modern Cultural Anthropology.* New York: Alfred A. Knopf.

Bröckling, U. (1997). *Disziplin. Soziologie und Geschichte militärischer Gehorsamsproduktion.* München: Fink.

Euskirchen, M. (2005a). *Militärrituale.* Köln: PapyRossa.

Euskirchen, M. (2005b). *Unter Kommando.* Retrieved June 8, 2005 from www.jungleworld.com/seiten/2005/20/5526.php.

Oxford English Dictionary (2008). *Ritual*. Oxford University Press. Retrieved February 22, 2008 from http://www.oed.com/view/Entry/166369.

Schmid, J. & Kocher-Schmid, C. (1992). *Söhne des Krokodils. Männerhausrituale und Initiation in Yensan, Zentral-Iatmul, East Sepik Province, Papua New Guinea.* Basel: Basler Beiträge zur Ethnologie.

Schweizer Armee (2004). *Dienstreglement 04 (DR04).* Bern: Bundesamt für Bauten und Logistik.

Streck, B. [Ed.] (2000). *Wörterbuch der Ethnologie.* Wuppertal: Hammer.

Studer, R. (2008, February 4). Im Abfallsack durch den eiskalten Bach. *Basler Zeitung*, p. 5.

Turner, V. (1964). Betwixt and between: the liminal period in rites de passage. In V. Turner (Ed.), *The forest of symbols* (pp. 23-59). Ithaca: Cornell University Press.

Turner, V. (1969). *The Ritual Process: Structure and Antistructure.* Chicago: Aldine.

Van Gennep, A. (1986). *Übergangsriten (Les rites de passage).* Frankfurt, New York: Campus. (Original work published in 1909).

Winslow, D. (1999). Rites of Passage and Group Bonding in the Canadian Airborne. *Armed Forces and Society, 25* (3), 430-457.

David Whetham

The Quest for Truth: What is a Credible Academic Source for a Defence Studies Essay?

For any academically assessed piece of work to be regarded as effective or deserving of credit, the argument will need to be supported by evidence. The important issue of plagiarism aside, citations are required so that the reader and marker can see where the information came from and make a judgment as to the quality of that evidence. Can the source be trusted? Is it credible? To what extent can it be said that the evidence presented is true? Within this context, objectivity is something that is often mentioned as a standard by which evidence can be judged. But what actually is objectivity or objective knowledge and why would we want it anyway? Is there even such a thing as objective knowledge; does it even exist and, if it does, is it possible to find it? Clearly, people generally perform tasks better if they are aware of what is expected of them and against which criteria they are being assessed. This paper is designed to provide some guidance for students of Defence Studies, but the ideas contained here are relevant across any academic discipline.[1] It will argue that while in one sense, the search for genuine 'objective knowledge' is bound to lead to disappointment, the process of seeking to acquire it is still a worthwhile and essential one and it is on this basis that some sources of information and evidence are to be preferred over others.

It is always useful to begin by defining terms. According to Collins English Dictionary (1994), to be objective is to exist independently of the mind. It is to be undistorted by personal feelings or bias and relates to facts rather than thought or opinion. It is that which is real. Conversely, that which is subjective is of or based on a person's emotions or prejudices (ibid.). Following from this, objective knowledge will be an unbiased appraisal of the world as it actually is. In the field of historical and social sciences, and any other academic discipline, objectivity is invariably preferred over subjectivity. While it may be fascinating to know what a particular individual thinks about a given topic, it is unlikely to make much impact on an academic debate if a contribution is limited to personal or anecdotal experience. Without reference to what other informed people have thought about the same area, or lacking legitimate evidence, arguments are unlikely

---

[1] Defence Studies is a multi-disciplinary approach to the study of defence. Students will not be trained as specialist historians, philosophers or social scientists, but they will be introduced to those aspects of their disciplines which are most germane to a rounded understanding of defence issues.

to be convincing. Personal musings may of course be extremely valuable as potentially one of many sources for researchers to build up a coherent picture, but the lack of the "broader picture" within it means that the work lacks the essential authority required for a genuine contribution to an academic discipline. But why should that be?

It is, perhaps, useful at this point to look at the very origins of epistemic enquiry. One of Plato's many concerns was with what a person can justifiably claim to know. He discusses on what grounds a statement might be established or refuted, and the types of argument that can be given for and against something (Hamilton & Cairns, 1989).[2] The *Theaetetus* in particular is concerned with the idea of knowledge (ibid.). It starts from the strongly empiricist claim (and common-sense position) that knowledge comes to us from the external world through our senses.[3] This naturally leads to the identification of knowledge with perception. While the technicalities of the Platonic position regarding perception are not of immediate interest here, the result of this is that Socrates (present in the dialogue) argues dialectically that as different people will perceive the same thing in different ways, "man is the measure of all things". This is the Protagorian doctrine of relativism and has important implications for any discussion of what constitutes knowledge and its relationship with objectivity. This ties in very clearly with many current debates as to why one view should be preferred over another: why should the BBC News be a more credible source of information than a fundamentalist Islamic website or the Project for a New American Century? Why should the *opinion* of a respected academic authority be taken as any better than the "man in the pub"? It is not even clear that we can all agree on what truth is anyway. For example, Bird (2008) quotes cross-cultural communication consultant Richard D. Lewis:

> It is very dangerous to assume that your foreign colleagues really understand what you mean when you use abstract and culture-bound terms. In Germany and Finland truth is scientific truth. In the Balkans, it is flexible. In Japan, it is a dangerous concept. In China, there is no absolute truth. (p. 50)

While intercultural differences can perhaps be overplayed, in a similar way, Post-Positivists argue that as there is no genuinely neutral position from which to judge the validity of one opinion over another, there can be no genuine objective truth; *all* truth is relative.[4]

---

2   Plato's early works were obviously strongly influenced by Socrates, who had argued so effectively against peoples' knowledge claims that he was profoundly sceptical as to whether even he could be said to actually *know* anything. See Plato, *Meno* 80, 86.
3   This view is also known as Positivism.
4   One of the most influential thinkers in this area is Foucault who asks 'how can history have a truth if truth has a history'. Truth is not something external to social settings but is instead part of them. See Foucault, 1984, 1994.

There are many problems with relativism, one of which is that it denies the possibility of meaningful discussion. All frames of reference become meaningless as they are only relevant for specific individuals and one is left in a hopelessly solipsistic state where nothing can be judged. For example, in the moral sphere, "It is wrong to kill" becomes simply the assertion that "I, personally, believe it is wrong to kill".[5] The premise of this paper is that some opinions and sources *are* better than others, so what actually is it that makes one source of knowledge better than a mere opinion? Essentially, knowledge (or a good source of evidence for an argument) is that which can be favorably judged against the standard of objectivity. In the world of academia and science, theoretically, objective knowledge can be reached by following a respectable methodology usable (see below) by any competent investigator: it is the way the world is to *any* person. Subjective knowledge is closer to opinion because it arises through questionable methodology and reflects an answer to a problem from a particular point-of-view. This is the way the world is to a *specific* person. This view may be very useful in some contexts but it also has limitations.

If one is to avoid relativism, then one would presumably prefer to rely on knowledge that conforms to this objective criterion. Is this a problem? Well unfortunately, it is. While many theoreticians prefer to shy away from this, the actual meaning of "objectivity" is bound up with some complicated metaphysical debates. The 'straightforwardly ontological concept' of objectivity is that which actually exists independently of any knowledge, perception, conception or consciousness (Bell, 2010). This sounds very straightforward, but depending upon one's metaphysical beliefs, what actually falls into this category is open to debate. For example, mathematics, the foundation of modern scientific thought is of particular concern as numbers may not actually have any objective reality. Frege, for example, felt that mathematics could comprise of objective knowledge only if the numbers it refers to, the propositions it actually consists of, and the truth-values it aims at, are all mind-independent entities. If numbers themselves are actually merely conceptual tools that we use to describe our environment then the very basis of the rest of our knowledge about the world is hopelessly subjective. Only by accepting (the decidedly unreal) position proposed by Plato where numbers exist as things in their own right in the world of forms, can logic, arithmetic and science be considered as truly objective. Common to these various strands of thought is that certain things like numbers and the real metaphysical nature of the universe do exist, but unfortunately, we cannot actually know them directly but have to experience them (effectively interpret them) through our senses and critical faculties. We are condemned to perceive the world as

---

5   For a good discussion of moral relativism in its various forms see Wong, 1994.

merely shadows on a cave wall (Hamilton & Cairns, 1989). This is Plato's metaphor for the fact that we are incapable of seeing the world as it really is. We simply have to make the most of these images, accept our limitations and interpret the world as best we can, getting as close to the objective knowledge that forever lies just out of reach.

If observation is supposed to be totally objective and distinct from that which is observed, then the empiricist model of knowledge is clearly in deep trouble right from the start. So where does that leave us? Can we no longer call anything knowledge because it lacks this essential objectivity? Without objectivity, relativism quickly rears its head. As we saw above, the natural result of accepting the subjective nature of all academic enquiry is solipsism. Nothing meaningful can be said because everybody has a genuine case for arguing whatever they feel. Every source becomes equally valid as evidence for an argument. This is clearly not the case but how are you supposed to determine what is going to be more or less acceptable as a credible source?

Fortunately, there is a practical response to this problem that avoids the pitfalls of relativism. If we can never actually judge our knowledge directly against the world as it actually is beyond our perception of it, then objectivity *in its useful sense* must mean something different. Such thinking considers the objectivity of a judgment as, amongst other things: a function of its coherence with other judgments; its possession of warranted grounds; its acceptance within a community, scientific or otherwise; its conformity with the a priori rules that constitute understanding; its verifiability (or indeed, falsifiability), etc. (Bell, 2010). For something to constitute genuine objective knowledge, it must be rational, justifiable, coherent, communicable and intelligible. We can only judge our knowledge of the world by our actual experience of the world as we see it. Any perspective is going to be just that – a perspective and one can never totally remove one's own understanding of the world from a picture – if one did, then the picture would probably no longer mean anything anyway.[6]

We therefore have to accept that no source of evidence can ever achieve a position of metaphysical objectivity, but this does not stop us from attempting to pursue an ideal. To do this we employ methods such as those designed to minimize what is known as the Rosenthal effect where the expectations of the experimenter about the outcome of an experiment will unwittingly effect the actual outcome of

---

[6] This is illustrated beautifully by a thought-experiment devised by Nagel (1989). Here, the attempt to objectively find out what it is like to be a bat is doomed to failure because what we actually want to know is what it would be like for *us* to be a bat. There is a subtle but incredibly important difference. The information can only be made sense of when it is seen from our viewpoint (and there is no point in showing the bat what it can see!).

the procedure (minimization and standardization of contact between experimenter and the subjects, expectancy-control methods, the use of a large number of experimenters dealing with a small number of subjects, and taking into effect possible bias before the experiment takes place)(Martin, 1994). Popper suggests that free criticism is essential to the operation of an objective science and that scientists should present their findings in a form that allows others to test and evaluate their theories by the impartial arbitrator of experience (Popper, 1953).

When dealing with the problem of values as motivations for an agent's actions or an author's theoretical perspective, even this can be looked at objectively. In *The Possibility of Altruism* Thomas Nagel demonstrates how one can move towards a position of objectivity without losing the relevance of the subject. Nagel sets out conditions for an ethical system that relies on people moving away from the purely personal viewpoint *without losing that viewpoint.* While the essence of personal judgments, beliefs or attitudes is that they view the world from a specific vantage point, the impersonal view or standpoint provides a view of the world without giving one's location within it. This does not mean that anything is lost, however, because according to Nagel (1978):

> A complete impersonal description of the world will include a description of the person who is 'I' in the personal description, and will recast in impersonal terms everything that can be said about that individual in the first person. Thus the impersonal standpoint should be able to accommodate all phenomena describable from the personal standpoint, including facts about the subject himself. (p. 101)

This technique can be used beyond the moral sphere of reasoning and can be a useful tool for attempting to assess anything objectively. Nagel developed this idea further in his work "The View from Nowhere" (1989). To give an example of the way it can work: I might have a phobia of spiders, and I also may have a fear of nuclear war. In everyday life, it is the phobia of spiders that will dominate my behavior and my worries, while the fear of a nuclear war may be rarely considered. If one were to step back, away from the personal position, however, one quickly puts these two concerns into perspective. I do not have to go very far away from my own viewpoint to allow a reassessment of these priorities. While many people are afraid of spiders, *objectively* they do not compare with the horrors of a nuclear holocaust. The farther away from the personal standpoint that I move, the more this is put into real perspective.

How does one apply these ideas to selecting evidence and building an academic argument? Individual personal experience is by its nature anecdotal and subjective. If it is employed carefully and with an obvious appreciation of context and the broader picture it can be useful as an illustration or example. However, an argument that relies heavily on this type of evidence is unlikely to be convincing. Academic journals generally contain meticulously researched articles, written by researchers who have a deep knowledge of a specialized area. These will be peer reviewed by other experts and if an article is considered to be unsubstantiated or contain poor analysis, it will not be included. Once the information or argument is in the public domain, other experts will criticize and challenge the work, exposing it to Popper's criteria of being tested and open to falsification. Unlike a magazine or newspaper, academic journals (to some people's despair) do not exist to make money, but to distribute knowledge. This can be contrasted with an article written by a journalist that presents a simplified or surface knowledge of a complicated area. A journalist may well produce a thoughtful and insightful overview, but her job is to make a complicated subject accessible by summarizing and explaining what others do. The journalist (outside of specialist experts who concentrate on only one area) is not an expert in the subject itself. That is why it is often best to 'go to the horse's mouth' rather than rely on an interpretation.

One tool that illustrates a number of issues in this debate is the multilingual, web-based encyclopedia called Wikipedia, popular as a first 'port of call' for many people looking something up. As of May 2012, it had over 22 million articles in 285 languages, about a sixth of which were in English (2012). There is no doubting its popularity and it certainly provides a fascinating and easily accessible source of information. Unfortunately, in many ways, its very strengths are also the factors that make it unsatisfactory as an academic research tool. The second part of the name is obviously derived from encyclopedia, but the 'wiki' refers to its nature as a collaborative website, editable by all – representing what its strongest advocates claim is the way the internet should be. However, the lack of single editorship means there is no central oversight. Because there are no experts on content to make a final decision, as there would be with a traditional reference encyclopedia, every article is in a constant state of editorial flux. For example, biographical data about a person can be changed by anyone else even if it was originally entered or corrected by the person whom the article is actually about. Nothing is ever in its final form – putting it in a positive way, every article is in the process of *becoming* correct. Anybody can modify the text as many times as they like through the edit text function and this can be done at any time. This, say Wikipedia's fans, is exactly why it is such an exciting and current resource with many people constantly checking if an entry is correct. However, that also means that no entry is ever in a final form to provide a definitive reference.[7] It also means that this anarchic editorship can lead to a lack of priorities. This is intellectual

---

7 This means that, in spite of its limitations, if Wikipedia is cited as a reference, it is essential to ensure that a date of reference is given as anyone wishing to verify an entry later may find that the text has changed radically.

relativism taken to the extreme as every article is of equal value, whatever it is about, as there is no system for prioritizing anything. Effectively, consensus is valued above credentials.[8] While Wikipedia can be an interesting starting point for research in many instances (for example, some articles will cite sources for useful further research), one must be very careful about actually relying upon it. While some of the better Wikipedia articles really are written by experts and are well researched, well referenced and remain unedited for long periods of time, it is very hard to know which ones they are. Working out which entries can be considered authoritative requires a subject matter expert, or further research employing a more traditional academic source. These are just some of the reasons which undermine Wikipedia as a credible source for academic purposes.

In contrast to this, the Reuters news group operates under a system of Trust Principles aimed at preserving its independence, integrity and freedom from bias. From these, journalists draw the core values of the organization which include that they will; always hold accuracy sacrosanct; always correct an error openly; always guard from putting their opinion in a story or editorializing; never fabricate or plagiarize, and; never pay for a story or accept a bribe (Schlesinger, 2007a). The immediate actions of the organization on discovering that a Reuters photo journalist had published two photographs from Lebanon in 2006 that had been digitally altered illustrate this. Reuters' relationship with the journalist was immediately terminated and Reuters ensured that the results of the subsequent internal investigation were shared with the public. As a result, the editor who was responsible for the oversight of the pictures was also dismissed, the editorial processes were changed, and additional training has been invested in to prevent any reoccurrence (Schlesinger, 2007b). This kind of transparent process is precisely why people around the world know they can trust the information that they receive through this organization, no matter who within the organization has written the material itself.

What is clear from the examples above is that with all research, it is essential to retain a healthy sense of skepticism. In particular, students should be warned not to believe everything they see on the web – anyone can publish anything with few (if any) checks, balances or quality control if they chose to avoid the more reputable sites. Some professional-looking websites are complete hoaxes.[9] A simple rule is that the bigger and more well-known an organization is, the more it has to lose if it provides false information. However, it is essential to

---

8 There are content policies and other measures employed by Wikipedia to remedy some of these issues, but it is difficult to see how they can be resolved without fundamentally changing the nature of the project itself.

9 E.g., "History of a Victorian Era Robot" http://www.bigredhair.com/boilerplate/intro.html.

constantly remember that everyone has an axe to grind and will present or interpret 'facts' in a way that suits them or their beliefs. For example, if a study denies the existence of global warming, it is worth checking to see whether a major oil company paid for the survey! Apply the lessons of Nagel's "View from Nowhere" to stand back from the source you are looking at – why do they view the evidence in a particular way? As you move back from the 'facts on the page' you start to see the context within which they are being interpreted. This is true when considering any source of information, including academic books; it is essential to see where the argument is coming from. Just quoting the name of the person who's opinion you have paraphrased as evidence is enough to avoid charges of plagiarism, but it is not enough to demonstrate to the person reading the paper that you understand the context and perspective of the argument. For example, in international relations theory, a Liberal will have a different perspective to a Realist on many subjects and may view exactly the same 'fact' in a completely different way. One only has to see how different newspapers can report exactly the same event in different ways to see that this is true. Understanding the theoretical perspective of an author or source and being able to place it within the broader picture is essential to being able to employ their ideas as effective evidence for your own argument.[10] A good example of this would be explaining why Kenneth Waltz, a leading international relations scholar, predicted the imminent demise of NATO following the end of the Cold War (McCalla, 1996). Once that you appreciate he is coming from a Neo-Realist perspective, concerned with the structures of the international system, it becomes obvious that NATO no longer has a function when it is not opposed by another large scale military alliance, and this situation should obviously lead to its disbandment. Waltz was compelled by the logic of his belief system to reach this conclusion. Returning to the examples used at the start, it should now be clear how a fundamentalist Islamic website will interpret the same event in the real world in a very different way from someone representing the Neo Conservative Project for the New American Century. Both will be considered more subjective than an organization such as the BBC, which is institutionally, professionally and legally obliged to avoid partisanship.[11] The BBC is not objective in the sense that it provides metaphysical truth, but it can generally be relied upon, and has proved itself over

---

10 This is part of what Miller, Imrie & Cox (1998) refer to as 'deep learning'.
11 The BBC was regulated by 12 Governors from 1927 until the end of 2006, when they were replaced by a 12 member BBC Trust. The Governors were charged with upholding standards and defending the BBC from political and commercial pressures, reporting to licence payers and Parliament. The new BBC Trust has a similar mandate to oversee the running of the BBC in the public interest. See http://www.bbc.co.uk/bbctrust/.

many years, to be less subjective than some other sources of information. It has both authority and credibility when compared to many other sources of information.

In conclusion, the search for genuinely objective truth is bound to lead to disappointment in any discipline on the grounds that it is unattainable. However, that does not mean that all sources are equally valid for providing evidence for academic pieces of work. Applying critical faculties and seeking to understand the limitations, inherent bias, perspective and context of any source of information is an essential part of developing within this intellectual environment. Part of a good argument or essay comes from demonstrating an understanding of the types of evidence available and being able to justify your selection. Knowledge is useless unless it relates to us in some way, but it is also limited in value *if it only relates to us as specific individuals.* The key to genuine knowledge (and a good essay) is getting this balance right.

Bibliography

Bell, D. (2010). Objectivity. In J. Dancy & E. Sosa & M. Steup (Eds.), *A Companion to Epistemology* (pp. 559-562). Singapore: Blackwell.

Bird, T. (2008). Go Global. *Blue Wings, May*, 49-55.

Collins English Dictionary (1994). *Objective* (p. 581). London: HarperCollins.

Foucault, M. (1984). *The Foucault Reader.* New York: Pantheon Books.

Foucault, M. (1994). *Power.* New York: New Press.

Hamilton, E. & Cairns, H. (Eds.) (1989). *Plato: The Collected Dialogues.* Princeton: Princeton University Press.

Martin, M. & McIntyre, L. (Eds.) (1994). *Readings in the Philosophy of Social Science*. Cambridge, MA: MIT Press.

Martin, M. (1994). The Philosophical Importance of the Rosenthal Effect. In M. Martin & L. McIntyre (Eds.), *Readings in the Philosophy of Social Science* (pp. 585-596). Cambridge, MA: MIT Press.

McCalla R. (1996). NATO's Persistence after the Cold War. *International Organization, 50* (3), 445-475.

Miller, A., Imrie, B. & Cox, K. (1998). *Student Assessment in Higher Education.* London: Kogan Page.

Nagel, T. (1978). *The Possibility of Altruism.* Princeton: Princeton University Press.

Nagel, T. (1989). *The View from Nowhere.* Oxford: Oxford University Press.

Nagel, T. (1989). What Is It Like to Be a Bat? In J. White (Ed.), *Introduction to Philosophy* (pp. 208-216). St. Paul, MN: West Publishing Company.

Popper, K. (1953). The Sociology of Knowledge: A Critique. In P. Wiener (Ed.), *Readings in the Philosophy of Science* (pp. 357-366). New York: Charles Scribner & Sons.

Schlesinger, D. (2007a). A brief guide to Reuters values and standards. Retrieved from http://blogs.reuters.com/blog/archives/4326.

Schlesinger, D. (2007b). The use of Photoshop. Retrieved from http://blogs.reuters.com/blog/archives/4327.

Singer, P. (Ed.) (1994). *A Companion to Ethics.* Oxford: Blackwell.

Subjective (1994) *Collins English Dictionary* (p. 861). London: HarperCollins.

White, J. (Ed.) (1989). *Introduction to Philosophy.* St. Paul, MN: West Publishing Company.

Wikipedia (2012). Wikipedia. Retrieved from http://en.wikipedia.org/wiki/Wikipedia.

Wong, D. (1994). Relativism. In P. Singer (Ed.), *A Companion to Ethics* (pp. 442-450). Oxford: Blackwell.

Aki-Mauri Huhtinen

The Changing Leadership Culture in Western Military Organizations

Command or Control?

David Alberts and Richard Hayes (2003) write in their study "Power to the Edge. Command and Control (C2) in the Information Age" how the superiority of an organization of the information age is compared to an organization of the industrial time, hiding in the utilization of the scope of the network and its edges. The hierarchical organization model of the industrial time was based on a threefold division into threat to the thought or the enemy, one's own power and a management system. The writers see the organization of the industrial time to be hopelessly outdated on the battlefield of the information time, where the worldview serves as the metaphor of Darwin's "survival of the fittest": the quick eat the slow. The organization of the industrial time is too slow in adapting to a constant change in the environment. According to the writers, the technological Revolution in Military Affairs (RMA) has been a fact for a long time already, but the industrial thinking culture and set of values of the armed forces are an obstacle for utilizing the technology (Alberts & Hayes, 2003; see also Doz & Kosonen, 2007).

The "bible of network centric warfare" (Alberts & Hayes, 2003) includes a distillation of the essence of command and control, providing definitions and identifying the enduring functions that must be performed in any military operation. Since there is no single approach to C2 that has yet to prove suitable for all purposes and situations, militaries throughout history have employed a variety of approaches to commanding and controlling their forces. A representative sample of the most successful of these approaches is reviewed and their implications are discussed. The authors then examine the nature of industrial age militaries, their inherent properties, and their inability to develop the level of interoperability and agility needed in the information age. The industrial age has had a profound effect on the nature and the conduct of warfare and on military organizations. A discussion of the characteristics of industrial age militaries and command and control is used to set the stage for an examination of their suitability for information age missions and environments. The nature of the changes associated with information age technologies and the desired characteristics of information age militaries, particularly the command and control capabilities needed to meet the full spectrum of mission challenges, are introduced and discussed in detail. Two interrelated force characteristics that transcend any mission are of particular importance in the information age: interoperability and agility. Each of these key topics is treated in a separate chapter. The basic concepts necessary to understand power to the edge are then introduced. Finally, the *advantages of moving power from the centre to the edge and achieving control indirectly, rather than directly*, are discussed as they apply to both military organizations and the architectures and processes of the C4ISR systems that support them.

No middle class war

Increasingly in Europe, nation states have given up liability for military service as a duty of citizens. The popularity of military culture as part of one's own ordinary life has decreased among the well-to-do middle class. War has an effect on both rich and poor people, and we once again have professional army systems like we had before the Peace of Westphalia in 1648. In regard to safety, the industrial time is being postponed at the information time. The role and organizational identity of the armed forces in the international global security become even more professional because the western armed forces become organizations equipped with high technology. In addition to cooperation and security and safety development, there is also the danger that armed forces become elitist, politicized, privatized and, in the worst case, criminalized. The ultimate responsibility of decision-making has been left to system management.

Duties of a citizen in the broad social middle classes of western democracies have bound the whole nation to the serving in armed forces. Now, this tradition has been given away. The outsourcing leads to a chain of subcontractors, whose reliability in crises and wars is unclear. Instead, slum-dwellers will make up the majority of the urban population, and we are thus witnessing the rapid growth of a population outside state control, living in conditions half outside the law, in dire need of minimal forms of self-organization. They are incorporated into the global economy in numerous ways. They may also be the potential human resource warriors of the future.

Military research and development

There is a global attitude that Europeans have no military capability or that they are unwilling to use their forces. But this is not true, as the United States may be able to operate on its own in Bosnia, Kosovo, and Afghanistan, but still needs help from the Europeans to maintain peace (Cooper, 2003). It is true that most European countries are ending conscription and are concentrating on a more professional and more mobile force. But it will take some time before the fruits of these changes become reality. Meanwhile, the United States is transforming its methods of military operation even more quickly. Quite soon even Britain and France, the most capable of the Europeans, may have difficulty operating with their US allies in the so-called digital battle space. In a technological age, military spending on research and development (R&D) and on equipment gives the best indication of capability.

Table 1. Military spending on R&D and equipment in $ millions in 2001

Country	R&D ($ millions)	Equipment ($ millions)	Total ($ millions) and relationship
Ireland	0	50	50
Denmark	1	224	225/ 0.000446
Austria	10	323	333/ 0.0301
Finland	8	618	626/ 0.0129
Spain	174	1.062	1 236/ 0.16
Sweden	103	2.114	2 217/ 0.048
Germany	1.286	3.389	4 677/ 0.37
UK	3.986	8.597	12 586/ *0.46*
USA	39.340	59.878	99 218/ *0.65*

(Cooper, 2003, p. 158).

As we see in Table 1, the relationship between R&D and military equipment is the most harmonized in the UK and the USA. In the European countries, the funding of R&D is weak because the lack of military capability encourages European countries to seek non-military solutions. Good equipment is critical to military success, but the use of force is not just about equipment. To be effective, European forces would have to develop experience in operating together. There is a profound asymmetry in European and American attitudes to the idea of a common defense. Many countries send their best officers to serve in the NATO headquarters at Mons, and most organize their military around NATO concepts, standards, and procedures, e.g., Finland. Europeans may be capable of territorial defense, but that is increasingly irrelevant in today's world. In many cases of the new threats, homeland defense begins abroad – in areas like Afghanistan (drugs), Balkans (human traffic) or Iraq (terrorism) (Cooper, 2003).

In the post-Cold War world, threats are much harder to identify and analyze. Every western nation state supposes that global problems, development crises, and regional conflicts have become increasingly significant for security. Along with globalization, nation states' internal and external security has become increasingly dependent on the broad international situation (Smith, 2007). Security threats and challenges are increasingly cross-border in nature. In responding to them, therefore, it is crucial to increase bilateral and multilateral cooperation in neighborhood relations, both regionally and globally, and establish procedures that are legally binding. For example, the most important point for Finland in this context is the capability and influence of the European Union. The role of the United States, the development of the transatlantic relationship and NATO's role and activities are also of key importance. After the general elections in the spring of 2007, the Finnish media emphasized the importance of the foreign political relations of Finland and the United States. Furthermore, the emphasis on the international relations between the armed forces is due to the concentration and expensiveness of the military-industrial complex. Small countries cannot afford their armed forces the time to develop the information any more.

The effects of the economy on the change in the organization structure of the armed forces are seen mainly inside the organizations and in the media (Gray, 2005). In contrast, the values of the armed forces of the traditional industrial time are further emphasized in the politicians' and senior officers' statements. The new needs of western armed forces have increasingly been created by the military-industrial complex. For example, the increasing trend towards non- or less lethal weapons does not stem from the nature of warfare, but from the military-industrial complex' way of finding a new way to make money.

In a new management or command and control structure, technology can be used to reduce the human resources of middle management. From the industrial perspective, it is fitting that traditional military organizations with vast human resources have to be reduced and transformed into agile and technologically flexible forces. The information to the organization of the time is indeed typical, a transition from the tripartite (strategical, operational, tactical) management structure to the two-party (strategical, tactical) structure. The core activity of armed forces has been based on the hiding of their own ways of action from the opponent. The capacity or efficiency of a nation state's military force has not been open to the global audience.

From the threats to the possibilities

After the Second World War, the development of western armed forces involved material and technological solutions to which the small countries had to adapt their own ways of action. Under the threat of nuclear weapons in the Cold War, one's own operation was regarded as well restricted (Ricigliano & Allen, 2006).

Today, cutting edge technology will define the conditions of the operation (Gray, 2005; Smith, 2007). When we develop the effect based organization, we can see the reality as different kinds of possibilities between increasing probabilities and impacts. Instead, in the classical situation we based the development of the military organization on different kinds of threats intended at decreasing probabilities and impacts.

Figure 1: The difference between effect based and threat based thinking.

The Difference between Effect Based and Threat Based Ideology

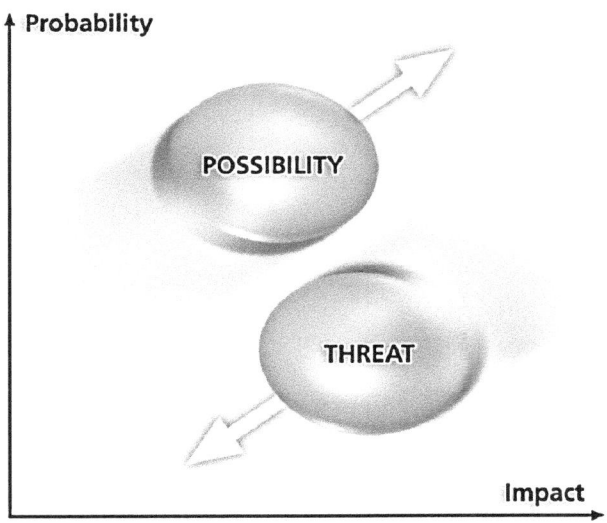

There is a great transformation going on between the concept of possibility and threat within the information age's security understanding. With the technology, threats are even desired, so that the significance of the technology remains high. Before, safety was based only on the threats as a probability and attempt were made to reduce their effect.

The war on terror has lasted several years now and the events have had deep effects on the global economy, politics, and military relations (Visuri, 2007). In the United States it was thought that victory would be achieved with Special Forces and air strikes. It was thought that large armies would not be needed any more. The war has devolved to costly guerrilla warfare and the quick network central war has not brought a result. As the war in Iraq became asymmetric, General David Petraeus was called to bring the task to end. The soft methods required in the situation and the large group numbers are not available to Petraeus because the United States has committed itself to the RMA warfare.

Even though the casualties are not at the level of the Korean or Vietnam War, the economic costs of the war already exceed the costs of the wars in question. In the war between Israel and Hezbollah in 2006, the guerrillas of Hezbollah were able to sustain Israel's bombings and were able to retain their position. On the media front, Israel suffered a big defeat. The control of the media by the United States is no longer as convincing as earlier, either. Arab and Asian channels are challenging the western media in their effectiveness.

In regard to the economy, the United States finances its war with debt money and its bases are being closed and their troops are being withdrawn, among others, from Europe. According to Visuri (2007), the American taxpayers will not pay the defense of Europe alone, but instead require European states to join the wars in Asia. Because the Asian countries are able to produce cheap infantry and masses of troops, one can see that in the battles of the future, the infantry will experience a renaissance, especially when the battle is fought in the population centers, cities, and urban combat theatres.

To prepare a new kind of war

Against the background of the privatization of armed forces, there is the unstated belief of the limitation of future wars and of wars no longer being fought in western democracies. On this foundation, the armed forces can be reduced and become partially technological societies. The defense forces no longer belong as a task to the operation in wide areas against numerous invaders. From this starting point, the culture of armed forces is facing big challenges in small democratic countries where armed force is assigned to the citizens' participation in the public good. Are the professional and elitist forces suited as a tool of political power in these kind societies? One reason for outsourcing and privatizing public administration is the war against media and the battle of reputation. We can also see a revolving door phenomenon, where the private sector nestles itself between citizens and the authorities of public service. For example, in the United States, former army officers can quite naturally continue their career in the business world.

"Planning early, planning twice" is still a relevant idea of warfare. According to General Rupert Smith (2007), battle is an event of circumstance, no matter how much planning, exercising, and drill precede it. The chance of victory is undoubtedly increased with proper preparation, but ultimately opponents fight the battle of the day: on another day, in the same location, with exactly the same forces, they would fight another battle in the different circumstances. The organization for the battle must always adapt. Command in war is exercised in the face of an opponent and in adversity. So is leadership. Today, we think that the opponent is like a passive "target" at the mercy of our superior technology.

Instead of planning, the military organizations have to change their action culture or praxis to living with planning. Usually, the high command and staff plan without living in battlespace. When the plan is ready, they try to adapt this plan in reality. But the real situation has changed during the planning process. In the information era there is no possibility to plan something outside of reality (Doz & Kosonen, 2007).

On a strategical level, every state needs three different kinds of skills: competence in actions, understandable communication systems, and the ability to translate global aims in specific local needs and values. In war as in the praxis of civilian organizations, the aim is the ability to fix an arbitrary line between domestic propaganda and foreign psychological warfare. For example, in Vietnam, America didn't have one version of the truth to give to foreign newsmen and another one to give to their American citizens (Kodosky, 2007).

To use another example, the Hezbollah is a single entity with a dual identity, i.e., a radical Islamic terrorist organization and a political party (Reich, 1990). It is both the long arm of Iran as well as a Syrian proxy. Its logic of thinking is based on terrorism and its concept of operation on guerrilla warfare. But the structure of the organization appears to be military. The strategic aim of the Hezbollah is victory through non-defeat. The operational-tactical praxis is based on continuous rocket attacks against the Israel Defense Forces (IDF) and Israeli cities. "Hiding – firing – hiding" as a way of waging war created an enormous challenge for the IDF. The Hezbollah used, e.g., the UN troops as shields for its operational preparations. The Hezbollah is an asymmetrical adversary and a postmodern terror organization with military capability.

The information age war is no longer an industrial war: the enemies are no longer the Third Reich or Japan who posed absolute and clear threats in recognizable groupings, and therefore provided stable political context for operations. As we have seen, our opponents are formless, and their leaders and operatives are outside the structures in which we order the world and society. The new threats are not directed against the state infrastructure or territories, but against the security of people, especially the psychological security of people (Smith, 2007; see also March & Weil, 2005).

There is a great advantage presently expected from the "digitization of the battlefield" and "network-enabled warfare". There is still the challenge of seeking to gain this advantage and the challenge of what it is sought over. There is a danger of knowing more and more about oneself and less and less about the enemy. The increase in information collection will probably be marked by a decrease in those elements that strike on the basis of the information gained, whether they are infantry, artillery, fighter-bombers, or warships. Their weapons and techniques will become more sophisticated (Smith, 2007).

The global change of military infrastructure in Western nation states can be represented as a deterministic and linear evolution. The high-tech weapons systems force the small Western nation states to transform their defense forces to similar systems and infrastructures as the United States has. It is a fact that today's "post-industrial" society needs fewer and fewer workers to reproduce it. This is the main reason why we can find only a security discussion on effectiveness and not on threats anymore (Ahern, 2007).

Many small western nation states have the same situation: you can use your money with material delivery, but there are not enough human resources to implement the new technology in a practical situation. We might speak of an economical "Trojan horse" of post-industrial military transformation.

The traditional small European armed forces based on the training culture are faced with a difficult situation due to the projects of information management and the C2 sector. The know-how of the organization and the turnover of the staff are not enough to get projects moving in the required direction. Because the European military culture is strongly based on national defense, the integration of the technological part of a armed force is culturally difficult. The plans of management and communication are often missing from the projects, in which case the know-how of the organization is depicted and communicated to the citizens merely in regard to its technology.

The system adaptation of the appropriate military leadership culture

Aidar (2002, p. 164) asked:

> As a strategic leader your prime responsibility is to ensure that your organization is going in the right direction. That sounds simple enough, but it is not always easy to achieve. What is the right strategic direction? How or where do you establish it? Why is implementation so difficult?

Aristotle has stated that practical wisdom, i.e., *phronesis*, is the central content of especially the officers', doctors' and lawyers' profession. Practical wisdom to make a forced decision is an ability in the conditions in which no sure mental starting point exists. This practical wisdom has not changed since Aristotle's times, while Western theoretical thinking and information changes its form continuously with an increasingly accelerating speed. Often the reality is the product of the human being and the abstraction which has come off a practical wisdom in many respects according to the central philosophers of the 1900s.

Many times we confound the planning process and thinking process. For example, strategic thinking is a function of practical wisdom, which is neither an art nor a science nor a skill. We forget that strategy really does not exist. We cannot stop or draw some strategy in the text on paper. We have to live and act within the strategy.

In ancient Greece the concept of strategy meant the whole art of a commander-in-chief, including leadership, administration, and working with allies, as well as knowing how to bring an enemy to battle and what tactics to employ. As armies became larger and warfare more complex, strategy was introduced as a new concept in contrast to tactics (Aidar, 2002). The ancient Greek concept of strategy also included the tactics as practical action levels. Now the concept of strategy has become increasingly abstract and the model of elitist process. Still, mental alertness, problem-solving ability and keen perception of relationships are all implicit in intelligence. The concept of *phronesis*, practical wisdom, means action that is the outcome of wisdom gained by experience (ibid.).

In his book "The Concept of Corporate Strategy" (1971), Professor Kenneth Andrews made an important contribution to understanding and improving the process of strategic thinking that precedes any form of planning. He advocated what became known as the SWOT analysis – that organizations should carefully appraise their strengths, weaknesses, opportunities and threats as a prelude to corporate strategy. He also stressed the importance of scanning the changing environment, using the ready-made headings of the political, economical, social and technological environment (PEST) to consider the salient factors. The pattern of decisions essentially means a plan (Aidar, 2002).

In the Finnish culture of the command and control process it has been customary to combine planning, education, and management. This way one and the same officer can be responsible for all the dimensions on his own organization level.

In US and NATO operations the planning is performed so that it is not connected to education and management of the practice. Different persons are responsible for education, the make-up of the groups, and management.

Figure 2. The concept of choice and change in strategy context.

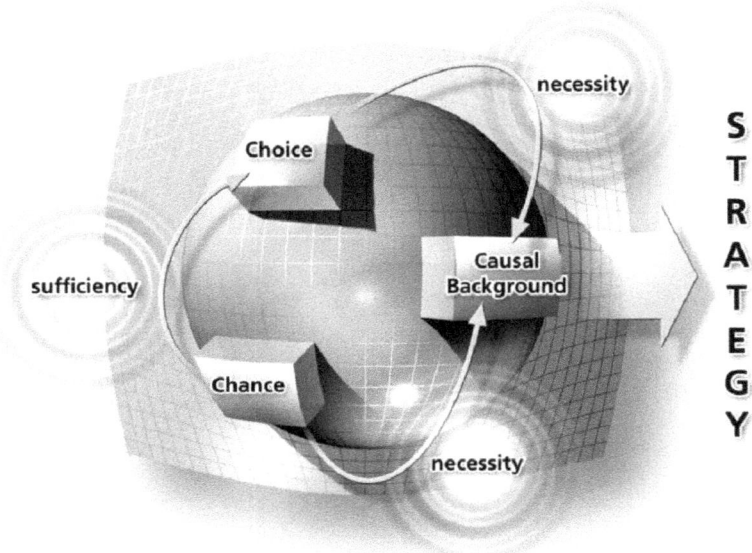

In strategy, rationality combines with intuition, chance, and a myriad of processes in which internal and external agents act, interact, tinker, and hesitate, taking advantage of some opportunities while failing to spot others (see Fig. 2). Formality, structure, and control are confronted with the informal, non-structured, and autonomous. In the former, decisions follow an orderly progression of problem identification, the search for solutions, selection, and implementation. In the latter, strategic choices originate from organizational garbage cans in which problems are generated from inside and outside the organization, and solutions are the outcome of random and opportunistic processes between actors. The relation between subunits and higher-level (or population level) units is deterministic, where population-level forces provide a comprehensive account of the behavior of individual units, leaving no scope for choice. By contrast, when subunits are different from higher-level units, yet related to them, the relation is one of heterogeneity (De Rond & Thietant, 2007).

The main question is where we start the investigation of strategy. According to Tolstoy, to study the laws of history we must completely change the subject of our observation, and leave aside kings, ministers, and generals, and study the common, infinitesimally small elements by which the masses are moved. First, strategy is seen to emerge from multiple, complex, interacting processes, only some of which are under managerial control. We have learnt to think that history obeys certain laws and the only thing is to discover its purpose. That is the reason why we find it so difficult to concentrate on the presence and being. The axiom that everything has a cause is a condition of our capacity to understand what is going on around us. Thus, throughout human history we have found it meaningful, even necessary, to think of events as somehow interconnected, as contributing to a grand, logical purposeful plot. Likewise, strategy, by most definitions, is naturally teleological in focusing on the means to an end (ibid.).

Managerial behavior is ordinarily directed towards the achievement of goals, intentions, or objects and the social sciences have generally concentrated principally on teleological determination. By contrast, natural science is primarily interested in efficient causes: formal and final causes are considered not amenable to experimentation, and material causes are taken for granted in natural phenomena. The strategic choices are fixed by the laws of nature and events in the distant past; given that it is not up to us what these laws and past events are, our choices are fixed by circumstances outside of our control. Hence, we are not free. By implication, we cannot be held accountable. If we are a product of a Darwinian evolutionary process, we can still choose the place, velocity and time of action. Still, the causal background is not in and of itself sufficient to produce strategic choice (ibid.).

The organizational actors need not only regard themselves as free but also need the concept of causation to be free. Causation as a strategic choice is implicated in a relationship of necessity; genuine freedom of choice cannot exist without presupposing causation. But, unlike the determinist, these causes are not in and of themselves sufficient to bring about strategy. Choosing requires deliberation. Hence, determinists and libertarians alike accept the presence of a causal background. But whereas the determinist will find their presence sufficient to account for particular choices, the libertarian insists on a gap between these and deciding. Where causal background is sufficient to determine a particular outcome we speak of strategic inevitability. The strategic choice can only ever be understood in terms of its relevant social and material context. A thing can only be homogeneous, heterogeneous, or independent with respect to something else. Causal background can be understood as the social and material context for decisions (ibid.).

The organizational structure – "the network-centric commander"

When we look at Western military history, we find that it does not matter what formation the military unit or organization starts in – square, diamond, arrow, line, column, or squashing matchbox – it will always end up in the same formation – small groups rallying around the bravest men or natural leaders (Aidar, 2002).

Alberts and Hayes (2003) promote a radical change of view in the art of war, the soldier organization, and its management. According to the writers, the interaction between the individuals and an organization requires new processes. They claim that the basic task of the organization of the industrial time was to serve the leader and that the information exchange and communication of the organization aimed at serving the decision-makers. This view of command could be characterized as power to centre, although the information age command can be characterized as power to edge. It is a shared and distributed responsibility. What, in fact, does being in charge mean in the networked warfare?

The writers establish their whole thought in that the notion of the undivided responsibility as a starting point for the soldier management is no longer true. The division of the responsibility to more actors is justified just in the utilizing of the variety of the network. Putting someone in charge does not result in effective command and control, but it makes the question of responsibility absolutely clear (Alberts & Hayes, 2003).

In 1963, Stanley Kubrick directed the classic satire *Dr. Strangelove Or How I Learned To Stop Worrying And Love The Bomb* with Peter Sellers and Georg C. Scott in the main roles. The film describes aptly a situation in which an accident caused by a human being happens to the management system, ultimately leading to nuclear war.

The film satirizes a delicate section of information management. The indirect management which takes place in the networks always contains a danger that the responsibility and the control of the operation slips from the hands of the management. In the film, the president of the United States repeatedly asks his generals how it is possible that the B-52 bombers enter Soviet air space even though he has not given such an order. The divided leadership is a target of the parody of the whole film. Eventually, the president of the United States has a personal telephone conversation with the chairman of the Soviet Union in order to prevent the nuclear annihilation of the world. Thus the film underlines the fact that, in the end, the responsibility for human lives remains undivided.

We are all familiar with the inability of economists to predict economic performance and the lackluster track record of various attempts to control the economy. All of us are familiar with efforts by meteorologists to forecast just one day into the future. The information age, according to Alberts and Hayes (2003), will separate the commander(s) from the function of command because commanders perform a variety of functions. This means that command and control processes no longer seek to optimize, but try to keep a situation within bounds while accomplishing an objective. Because of the media and the individualization of

Western society, risk management and protection of one's own force has taken an increasingly important role for defense forces. The cost of each single soldier has changed. In the information age, control can only be achieved indirectly. Control is not achieved by imposing a parallel process, but rather emerges from influencing the behaviors of independent agents. Instead of being in control, the enterprise creates the conditions that are likely to give rise to the behaviors that are desired.

Table 2. The difference between hierarchies and edge

	Hierarchies	**Edge**
Command	By directive	Establishing conditions
Leadership	By position	By competition
Control	By direction	An emergent property
Decision making	Line function	Everyone's job
Individual	Constrained	Empowered
Information	Hoarded	Shared

(Alberts & Hayes, 2003, p. 220)

Today, the network connects young people, but at the same time separates the ones belonging to different social classes into their own pigeonholes. Instead of an income level, race, or sex, seeking a community where a young person knows he or she belongs now has an even greater effect. The network is a question of identity. In addition to the advertisers, the division is utilized, e.g., by the United States Army, which closed the access to Myspace (favored by the soldiers) but left it open to Facebook (favored by the officers) (Boyd, 2007).

Still, the military ranks and the resulting control of the soldier organization cause authority conflicts, even though it has moved to a netlike organization structure. In the networks, information and tasks change place from one expert to another and often the information does not pass the superior because he does not belong to the expert network anymore. The network as a technological solution has broken the forming of management culture and planning, an executive group separated from each other.

The leaders of the so called line-and-staff organization will experience a defeat of authority when the subordinates network according to practical needs. We notice that from the point of view of safety, to give birth to information and understanding is more important than an ability to make quick decisions. The speed of decision heuristics fascinates us but at the same time leads to the constant inflation of the making of decisions. The ability of decision making is a play.

Earlier one could live in a situation in which the leaders' decision preceded planning. The subordinate experts were allowed to look at the justifications for decisions made afterwards in the management teams. Now the junctions will dictate the forming of the snapshot in the networks to the planning and decision-making in chronological order.

Strategic thinking should differ from strategic planning, which gives the model for the decision-making. However, the administration of the security of countries is a culture where a small core group often makes a decision before having scientifically corroborated results on a matter or phenomenon and also before the administrative planning. The role of the scientific study in the administration of security is indeed to look for the grounds of already made decisions.

However, the globalization and the networking of environments are changing this traditional order of management. Now planning is a foundation for the learning and informing of the whole organization. Complexity of the situations and an infinite amount of information compel to give up an ability of decision making as a starting point for everything and to move the focus of the management for the success of communication and interaction. The global media and the different public groups react in real-time to the decisions made by the organizations and to their practical consequences. For the organizations, management of the reputation has indeed risen past the ability of decision making to the centre of management.

Figure 3. The influence of changing environment to management and leadership.

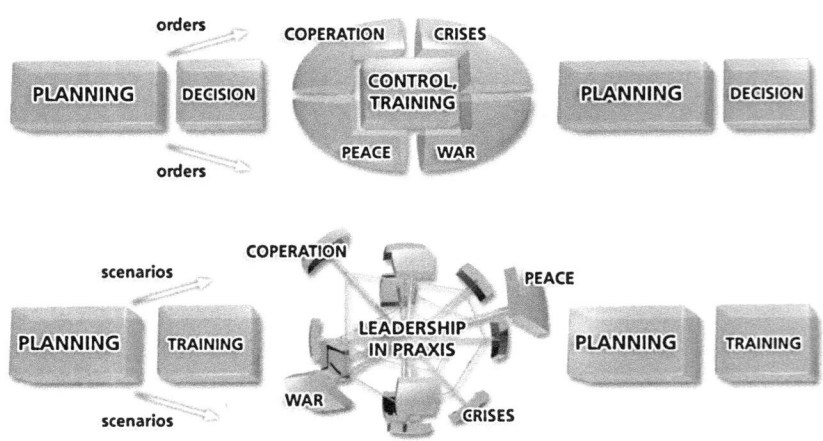

According to the General Field Doctrine of the FDF (Kenttäohjesääntö, 2007, p. 32), leadership in all situations is based on a model of the line-and-staff organization, and tasks are bound to the military rank, rights, and responsibility of everyone in the organization structure in all the situations. In addition to the FDF, the principle of performance management is also used and different phenomena and matters, such as the delivery of a weapon system, are carried out with the methods of process management. In practice, many processes, such as the delivery of a weapon system, begin with a clear ownership, but in a line-and-staff organization the tasks also often change in the organization and the owners of the process will change. In the line-and-staff organization the power and the responsibility are bound to a hierarchical and bureaucratic position whereas in a process responsibility is based on the time span of the project. When an officer owns processes and changes tasks in the line-and-staff organization, verifying power and responsibility issues will be difficult. In principle, simultaneous process organization and matrix organization which functions over line-and-staff organization possibly make the question of power and responsibility less pronounced.

The overlapping of two different management methods is manifested as a challenge especially in the acquisition of weapon systems and military material. The high-tech weapon systems of the information age and their possibilities of use are based on complex networks of studies, development, and the international court, in which the time span required for carrying out a project is considerably longer than the turnover of the staff of the line-and-staff organization. When an officer is in the situation "sitting on two chairs", the power and the responsibility issues will be very complex. One might ask if the officer changes the station and the task in the line-and-staff organization during the management of the project, according to which management structure will the officer's power and responsibility be estimated? How will the quality of the process be secured when its owner changes? Furthermore, more and more outside actors of the soldier organization and outsourced subprocesses often participate in the process. In this area power and responsibility issues often depend on the agreements which are based on the law, and the combining of the ways of action of different organizational cultures has not yet taken place.

The existence of two overlapping structures makes advocating one's personal interests for an officer possible. A person who is unscrupulous and pursues his own interest can start new and interesting projects, but in the line-and-staff organization he can fall back on a protection, which is based on the position of withdrawing quickly when the project runs into difficulties. The next officer to lead the project may receive a task the grounds of which are unknown to him. For the new owner of the process in question, the possibility to succeed is nonexistent before he even begins. The failure of the next officer in the promotion of the process will weaken his status in line-and-staff organization.

Alongside traditional power and responsibility based on station and task, a horizontal leading of the project and process management based on professionalism takes place, giving birth to exceptionally strong competition and pursuit of one's own interest in the culture of the soldier organization. Passiveness of particularly the peacetime environment allures the competitive and victory-focused soldier culture to also utilize the ruthless pursuit of one's own interest. When furthermore our time favors the promotion of socially apt and well mannered people to positions of leadership, there is the danger that the overall interests of national security is endangered by individual and personal aims. The situation of two systems does not make the democratically required transparency of power and responsibility possible either.

The postmodern rhizomatic organization

The nervous system is unlike that of a conventional armed force. The conventional system is in essence hierarchical: information flows up from the bottom and orders and instructions flow down, being disaggregated into detailed tasks at each point in the chain. In this way the whole force is concentrated on achieving its singular military strategic objective, with every individual action and achievement contributing coherently towards that end. But the system is vulnerable to the loss of a point of command – in which case the chain is broken.

Modern communications have been applied to this basic model, but the model is still the foundation. Guerrilla, and in particular terrorist, nervous systems do not work in this way, mainly because of their dependence on the people and their lack of strategic military objectives. To use a botanical analogy, their nervous system is "rhizomatic". Rhizomatic plants can propagate themselves through their roots; nettles, brambles and most grasses do this. They can increase by spreading fertilized seed, or vegetatively through their root systems, even when the root is severed from the parent body. A rhizomatic command system operates with an apparently hierarchical system above ground. It is a horizontal system, with many discrete groups. It is visible in the operational and political level, but invisible in infrastructural level. The cells operate by getting others to do the dirty work, or indirectly through some front organization. In all cases the need for security is paramount. A cell will do a minimum of three things: direct and sometimes lead military action, collect and hold resources such as money and weapons, and direct and sometimes conduct political action. Different people usually carry out the different functions. The centre will reinforce successful cells with funds, skills and weapons, seeking to establish an area of sanctuary from which to develop. If the guerrilla fighter can show himself to the people as their defender, risking his life for the greater good of all, then they will support him. The rhizomatic command system is difficult to attack, just as rhizomatic weeds are difficult to eradicate. To cut and roll encourages the root system to spread and put forward more shoots. You can only dig them up by poisoning or removing the nutrients from the soil (Smith, 2007).

In the security culture decisions aren't networked or democratized. The logic and framing of a question and the time span of political, military and economic decision-making do not correspond to the time span of the scientific basic research. Instead, the study of the technology will be well successful to adapt to the future which is also suitable for economical and political decision-making.

The central question is indeed: which stage in the globalization and networking will break the station and have traditional authority to move the decision-making to the network? What does the change mean for the structures of the administration of security?

For example, according to Kesseli (2006), internationalization to some extent also confuses the definition of military culture in Finland. Because Finnish units are nowadays essential elements of multinational troops on different peacekeeping missions, we prepare to take part in Rapid Deployment Forces (RDF) in the European Union. Global cooperation in Multi National Experimentation (MNE) exercise has also started, so it is only natural that definitions of military art must be standardized. The first steps have been taken, including translating Guidance for Operational Planning (GOP) – including definitions – into Finnish (FINGOP). In addition to standardizing definitions, FINGOP has been trying to take into consideration local special demands as well. However, learning about a new culture does not occur overnight. This also poses – and has already posed – challenges to both the research and teaching of military art, as our traditional understanding of the levels of military art is changing, or at least we can say that the scale of military art is changing by virtue of internationalization. However, it also has to be said that somehow the Finnish practical way of tying the levels of military art to the size of the organizations or to the range of the areas of responsibility helps us to revise the definitions. It remains to be seen whether the definitions described above are enough, if ever-smaller and more scattered forces will have more and more challenging tasks on an ever-larger battlefield.

When at the same time the industry society becomes an information society, it will change from a society of threats into a society of capacities. New threats, like terrorism, organized crime, or illegal immigration call for new capabilities. The central foundation of the capacity of a defense system at the FDF is the citizens' national defense spirit. The citizens' values and traditions and the general liability to military service have been cornerstones of the national defense spirit. This spirit is based on principles, knowledgeable and motivated staff, and material (Kenttäohjesääntö, 2007).

The question of leadership has had a difficult history after the Second World War. Because of Hitler, Stalin, Mussolini, etc., the strong personality has been seen as a negative phenomenon. Anti-authoritarianism is becoming the main idea of leadership. We forget that the quality that makes people trust leaders depends on cultural environment and values situation.

In a classical situation, in a small unit, the leader should be able to do the job of any man in the outfit better. As the leader rises higher in the order, he can no longer be expected to show such mastery of the details of all activities under him. But he has got to know how long these jobs should take, what their difficulties are, what they need in training and equipment. As the leader moves towards the top of the ladder, he must be able to judge between experts and technicians and to use their advice although he will not need their knowledge (Aidar, 2002).

When a new weapon technology emerges, small countries must reduce the number of the groups to check costs. Traditional functions, such as logistics, care services, maintenance services, and measures of support must be outsourced. Business companies take care of the tasks of the armed forces of Western countries more and more frequently. How can the leader be sure about the reliability of the operation if subcontractors are used and the subcontractors move on the limits of agreements to maximize economic profits? How may the leader responsible for the core functions be able to follow the hierarchy and network formed by the subcontractors of several levels?

If we examine the structure of the organization of the Western armed forces in the light of the core functions and support functions, the organization will not form a clear threefold division into strategic, operative, and tactical level any more. The planning and decision-making in reality do not follow a classic threefold division either. Instead, the armed forces take part in the global network in which economical, political and media related agreement networks also affect the operation of armed forces. The actual agents are junctions of the network through which the critical information currents run. These junctions can also be on the strategical, operative and tactical level.

Let's take an example. According to Klare (2007), the average American soldier in Iraq and Afghanistan consumes sixteen gallons of oil on a daily basis either directly, through the use of Humvees, tanks, trucks, and helicopters, or indirectly, by calling in air strikes. Multiply this figure by 162'000 soldiers in Iraq, 24'000 in Afghanistan, and 30'000 in the surrounding region (including sailors aboard US warships in the Persian Gulf) and you arrive at approximately 3.5 million gallons of oil: the daily petroleum tab for US combat operations in the Middle

East war zone at the height of the operations. Multiply that daily tab by 365 and you get 1.3 billion gallons: the estimated annual oil expenditure for US combat operations in Southwest Asia.

Klare continues how for a superpower's high-tech army located halfway around the world, the US Department of Defense must move millions of tons of arms, ammunition, food, fuel, and equipment every year by plane or ship, consuming additional tanker-loads of petroleum. Add this to the tally and the Pentagon's war-related oil budget rises appreciably, though exactly how much we have no real way of knowing.

Possessing the world's largest fleet of modern aircraft, helicopters, ships, tanks, armored vehicles, and support systems – virtually all powered by oil – the Department of Defense (DoD) is, in fact, the world's leading consumer of petroleum. According to Klare, the Pentagon might consume as much as 340'000 barrels (14 million gallons) every day. This is greater than the total national consumption of Sweden or Switzerland.

Even if the superpower can in the near future find enough resources for maintaining its armed forces, how will small countries succeed? The armed forces and their newest equipment will still need fuel in the future. With regard to the tools of future warfare, one can return to the nuclear weapons and also to the guerrilla weapons of the low tech era.

At the moment in the West, the armed forces want to project a certain image outwards. For example, the entrance requirements for the armed forces are kept high on both physical and psychological level. The mere desire is not enough anymore, but one must be efficient. One can ask if, however, all the citizens are needed for the security of the country during a real danger. The question of the human being a father, husband, son, mother and citizen has been always buried under different systems. The speech from a division to a regiment, battalion or company buries the question of the human being. Correspondingly, the speech on the management or command and control system or the network from the defense forgets to mention the place of the human being.

Where the cyborg connects to the machine and adapts the human being, the suicide bomber adapts the human being to the ideology and religion. Since the year 1922, virtually every male Finn has been connected to the tradition of the national defense as a human being. A central question of solidarity during our time is: how do the groups of different identities experience that they belong together? And how does the Internet connect the subcultures to each other at the moment of the distress? How do we identify ourselves and our enemy if there are targets of the identity in several places? In the political as well as the economical and military dimension there are more and more subcultures with which the possibility to identify becomes more fragmented.

In the so called post-Cold War, the only reasonable way of finding identity is to constantly ask. Situational awareness means asking, not telling, because of constant information flow in networks.

It is not possible to speak about a learning organization in military academy culture because all cadets have to study the same things and mutual competition is the key element of the curriculum. How can we measure the competition in the organization in which the individual can freely choose their interest of study and education?

Each individual needs to know the strategic importance of what they are being asked to do and their precise part in the drama that is unfolding (Aidar, 2002). The leader and the men who follow him represent one of the oldest, most natural and most effective of all human relationships. The manager and those he manages are a later product, with neither a so romantic nor so inspiring a history. Leadership is of the spirit, compounded of personality and vision; its practice is an art. Management is more a matter of accurate calculation, of statistics, of methods, timetables and routine; its practice is a science (ibid.).

Human Resources are a much-used phrase these days, employed now as a substitute for the older (military) term personnel. It is a bit of management-speak really, for it appears to classify people as if they were economic resources, along with finance, machinery and energy. The phrase human resource really describes that inner or hidden reserve of energy, life or spirit that lies within the human individual (ibid.).

If we have a manager attitude, we often delegate the work that does not interest ourselves to other people. Delegating interesting and challenging work is a means of developing people.

In Finland young people get a direct benefit for service as conscripts. This does not mean the benefit of quick career development but preparing for the social skills in the working life. Especially the leader education acquired during the service is considered an advantage on the labor market. About 25'000 young people serve in the FDF in Finland every year; about 80% of the whole age group. So the liability to military service is culturally a socializing phenomenon in Finnish society.

The identity of the human being is based on the group. In Finland, the liability to military service have reached deep into the society and culture. The danger of outsourcing it risks destroying the identity and the national defense spirit. The savings will be gained from the human resources when they should be gained from technological acquisitions. We are too constrained to adapt our own operation to the conditions appointed by the technology.

Many times, behind the concept of security, there is an idea of normality, standard, and rule. The security architecture is based on a business risk model and on operational risk management. Rich interconnection in the target may produce a disproportionate scale of effects, e.g., viruses, social panic, loss of trust, common mode failure, saturation of capability, pollution, and financial failure. Real-time risk management will become essential. Time banding is relative because it takes months to create a strategy and it happens in a social context, when instead it takes only hours or minutes to create a task and it happens in a rational or cognitive context. We need to concentrate on so called web science (mathematics, computer science, web engineering, artificial intelligence, law, psychology, sociology, social-culture, media, and ecology). We also need to identity management and information pedagogy. According to Henry Giroux (2008), the growing influence of a military presence and ideology in American society is made visible, in part, by the fact that the United States now has more police, prisons, spies, weapons, and soldiers than at any other time in its history. Militarization is simultaneously a discursive process, involving a shift in general societal beliefs and values in ways necessary to legitimate the use of force, the organization of large standing armies and their leaders, and the higher taxes or tribute used to pay for them. Popular fears about domestic safety and internal threats accentuated by endless terror alerts have created a society that increasingly accepted the notion of a "war without limits" as a normal state of affairs.

According to Giroux, we can speak of new "military metaphysics" which mean to define international reality as basically military. It is a tendency to see international problems as military problems and discount the likelihood of finding a solution outside of military means. Militarism is a new epistemology defining what is fact and fiction, right and wrong, just and unjust. One element of militarism is the growing collapse of the separation between church and state, on the other hand, and the increasing use of religious rhetoric as a marker of political-economy identity and in the shaping of public policy, on the other. According to Giroux, official *newspeak* also trades in the rhetoric of fear in order to manipulate the public into a state of servile political dependency and unquestioning ideological support. Military language is used in this context to say one thing but to actually mean its opposite, as in George Orwell's dystopian world in 1984 (ibid.).

Conclusion – Toward Leadership Environment

Strategy, as a product of a mind or minds, once incarnated as a plan, is like any other artistic product – a book, or poem, a musical composition – in that it can take on a life of its own, quite independent of its creator. A plan is a very good basis for changing your mind. Many times we think that flexibility is a weakness.

We know from military history how, for example, the so called Schlieffen Plan was not successful because of its lack of flexibility. The idea behind the plan was great, but the time and political environment had changed, and so the idea and the plan were no longer in balance (Aidar, 2002).

It is difficult in military culture to understand that some new process or political change may have come along overnight and you have got to adjust yourself and your organization to it.

According to Gray (2005), four elements make up the climate of war: danger, exertion, uncertainty, and change. Information superiority should lessen the salience and effects of those elements, but it cannot eliminate them. Information and technological solutions sink no ship in battle space. In war and strategy there are many dimensions, of which intelligence, or information, is only one. No measure of information dominance will compensate for low morale, poor discipline, or inappropriate or otherwise inadequate training on the part of the troops at the sharp end of war. Similarly, information dominance will not rescue a military venture that is poorly conceived strategically or incompetently commanded. Warfare is just too richly textured, too multidimensional, to be reduced to decisive resolution by information superiority.

The cutting edge of cyber power is developed for commercial motives. In surface, information warfare or cyber warfare looks like bloodless warfare, because it is itself non-lethal. Because cyber warfare is instant warfare, it is also using the global communication networks. Cyber power is accessible to all, and that is why it is a joint team player or it is inconclusive. Cyber power, its strengths and weaknesses, has joined the arsenal of war on a permanent basis. Also there are no laws, rules, or norms providing governance for cyberspace. Without the increasing high-tech solution in warfare, a little information power goes a long way in military command chain.

War is organized violence threatened or waged for political purposes. If the behavior under scrutiny is other than just defined, it is not war.

I would like to summarize some key points of my arguments. First, strategy is a military concept of origin. Strategic thinking should be distinguished from strategic planning. Strategic thinking leads to strategic planning. Strategy is the art of the leader-in-chief. The practical wisdom – intelligence, experience, and goodness – is the basis of strategy. It cannot be taught like a science or skill, but like any art it can be learned by those who have an aptitude for it. Planning is a process, not a destination. The golden rule is flexibility of mind, so that you can adapt after the plan but still make forward progress as circumstances unfold. Vision is the art of seeing things invisible.

Second, the changing of thinking is central to leadership. The changing process is based on the understanding of culture. Culture is wider than behavior:

it embraces the distinctive customs, achievements, products, outlook, values, and beliefs of a society or group, and the way of life.

Third, the difference between leadership and management is that in leadership it is not enough to merely read a new order, but you have to also see it in praxis (Aidar, 2002). Only face-to-face communication can change the organizational culture. Culture is not the same as structure, which is relatively easy to change. It is the deeper-seated pattern – unique to every organization – of assumptions, beliefs, attitudes, habits, and customs (ibid.).

The living condition for the success of companies is the creation of their own databank with the help of their own study and development. In the FDF, the possibility to spend money on consultant services that began in the 1990s is a development path in which the following turn will be the creation of an own study and research culture. Instead of the consultants, researchers and information leaders are placed in their own organization. However, this change requires a new separation of powers and opening of operative decision-making for a wider discussion within the organization. The commander-centrality and the personifying of the omnipotence of the information are no longer suitable for leadership of the information time. How well networking and divided leadership will stay in crises remains to be seen when the abilities of the new management will be weighed in the future.

The philosophy of C2 information warfare is that in the industrial time, management came from the hierarchy and all honor, communication, and creativity were concentrated to one visible commander. Therefore the creativity and snapshot of the organization were always late, of course, not least because of a deficient technology. Now we are going through a time of leadership or C2 network centric warfare. In the networks every one is a leader and the era of one visible leader is over. While previously a leader engaged in the art and the staff in the science, it is now actually the other way round. But RMA leaves two factors open: First, how is responsibility specified in the networks? Who will be responsible in the ultimate situation if the technological system causes a catastrophe? Second, because ordering no longer proceeds top-down in the line, the commanding of the organization in the middle of all creative chaos requires indirect control: the end users (human beings) are influenced so that by changing their behavior they will be compelled, for example, to use information processing systems of a certain kind and mobile phones. The technical battle solutions direct people to learn and this way the self itself is controlled, in other words we can call it "development" as a human being. What happens to those people who do not learn or who are not able to use the available military technological solutions? What does it cause to the solidity in the group? Will there eventually no longer be people in the battle state? Will war not belong to the sphere of cultural history?

Bibliography

Ahern, S. (2007, March). *Learning during War? Analyzing Organizational Change in the Army's Post-9/11 Environment.* Paper presented at the 48th Annual Convention of the International Studies Association, Chicago, IL.

Aidar, J. (2002). *Effective Strategic Leadership.* London: Pan Books.

Alberts, D. & Hayes, R. (2003). *Power to the Edge. Command and Control in the Information Age.* CCRP Publication Series. Retrieved from http://www.dodccrp.org/files/Alberts_Power.pdf.

Andrews, K. (1971). *The Concept of Corporate Strategy.* Homewood: Dow Jones-Irwin.

Boyd, D. (2007, June 24). *Viewing American class divisions through Facebook and MySpace.* Message posted to http://www.danah.org/papers/essays/ClassDivisions.html.

Cooper, R. (2003). *The Breaking of Nations. Order and Chaos in the Twenty-first Century.* New York: Grove Press.

De Rond, M. & Thietant, R.-A. (2007). Choice, chance, and inevitability in strategy. *Strategic Management Journal, 28*, 535-551.

Doz, Y. & Kosonen, M. (2007). The New Deal at the Top. *Harvard Business Review, June*, 98-104.

Giroux, H. (2008). *Against the Terror of Neoliberalism. Politics Beyond the Age of Greed.* London: Paradigm Publishers.

Gray, C. (2005). *Another Bloody Century. Future Warfare.* London: Weidenfeld & Nicholson.

Guttieri, K. (2007, March). *Words as Weapons: Strategic Communication in the War on Terror.* Paper presented at the 48th Annual Convention of the International Studies Association, Chicago, IL.

Kenttäohjesääntö (2007). *Yleinen osa. Puolustusjärjestelmän toiminnan perusteet.* Pääesikunta, Suunnitteluosasto. Ohjesääntönumero 202. Helsinki: Edita.

Kesseli, P. (2006). War Studies at the Finnish National Defence University. In T. Kristiansen & J. Olsen (Eds.), *War Studies. Perspectives from the Baltic and Nordic War Colleges. Oslo Files on Defence and Security 02/2007* (pp. 63-85). Norwegian Institute for Defence Studies. Retrieved from http://brage.bibsys.no/fhs/bitstream/URN:NBN:no-bibsys_brage_26270/1/OF_2_2007.pdf.

Klare, M. (2007). *The Pentagon vs. Peak Oil. How Wars of the Future May Be Fought Just to Run the Machines That Fight Them.* Retrieved from http://www.zcommunications.org/the-pentagon-v-peak-oil-by-michael-t-klare.

Kodosky, R. (2007) *Psychological Operations American Style. The Joint United States Public Affairs Office, Vietnam and Beyond.* New York: Lexington Book, Rowman & Littlefield Publishers.

March, J. & Weil, T. (2005). *On Leadership.* Oxford: Blackwell Publishing.

Reich, W. (Ed.). (1990). *Orgins of Terrorism. Psychologies, Ideologies, Theologies, States of Mind.* Washington, D.C.: Woodrow Wilson Center Press.

Ricigliano, R. & Allen, M. (2006). Cold War Redux. In A. Martin & P. Petro (Eds.), *Rethinking Global Security. Media, Popular Culture, and the "War on Terror"* (pp. 85-103). London: Rutgers University Press.

Smith, R. (2007). *The Utility of Force. The Art of War in the Modern World.* New York: Alfred A. Knopf.

Visuri, P. (2007). Suurvaltojen sodankuvat ja Suomen näkökulma. *Kylkirauta, 2,* 5-11.

2. Military Pedagogic Teaching and Practice

Amira Raviv

Military Ethics and Moral Dilemmas: Between "On the Job Learning" and Formal Education

Overview: Ethics as the Moral Fiber of the Army

Ongoing concern with ethical and normative issues constitutes the moral and existential fiber of any army in a democratic country. Therefore, the development of a moral doctrine is the basis for shaping a separate, distinct identity that lends depth and validity to the military career. The development of the moral and ethical awareness of commanders is essential in an organization whose existence focuses on acts of violence and use of force, which are forbidden from a moral point of view.

However, moral identity, which is often expressed in codes of ethics, cannot be separated from the norms prevailing in the society and in the country at a given time[1]. Moreover, over the years the gap between the civil norms of Western societies and the value system of the army has grown. The erosion of values such as collective discipline, sacrificing one's life for the country, responsibility, personal example, and even identification with and legitimation of the goals of war appear to have influenced the army and penetrated into the military system. Erosion exacts a price which might affect the spirit of combat, the quality of performance in missions, and the norms of conduct in routine and emergency situations. However, the gap between civil and military norms also arouses welcome skepticism. Thus, there is a need to constantly examine the gaps and to seek bold responses to questions of identity.

This gap also explains some of the increasing difficulty that Western armies have encountered in the attempt to instill values and grapple with ethical dilemmas. However, a second reason for the complexity of instilling military ethics is the changing face of war, as reflected in the transition from high-intensity to low-intensity warfare. In that process, large-scale national war, which is characterized by space, mass, and organized and uniform forces, has been replaced by a struggle with a diverse range of threats emanating from various sources, e.g., esoteric, grass-roots civil movements spanning the globe, often engaging in subversive, gang-like activity. As a result of this transition, the military organization faces numerous and more complex dilemmas.

---

1   I would like to thank IDF Reserve General Roni Suleimani from the Education and Youth Corps for this important insight.

The Uniqueness and Complexity of the Israeli Case

Israel has been waging an ongoing struggle with terrorism. In that sense, Israel is the only democracy in the world that has been coping for years with various types of military threats, and particularly with the threat of terror. Unfortunately, the broad scope of these threats has provided opportunities to learn, understand, and cope with the situation, as well as to make decisions about military actions in general and complex ethical considerations in particular.

The threat of fanatical terror that Israel faces has taken on various forms over the years, including: Palestinian suicide bombings, where terrorists have been sent to carry out suicide attacks in the heart of Israeli cities and among Israel's civilian population; terrorism perpetrated by the Hezbollah, where hundreds of rockets have been fired indiscriminately on localities in the north of Israel; attacks from Hamas, where thousands of Qassam rockets have been fired on the town of Sderot and other cities in the southern region of Israel. This indiscriminate terror has forced the civilian population of Israel to deal with fear, destruction, and death.

The tactics used by the terror organizations are based on psychological warfare, where the residents' sense of personal security is constantly undermined, and deliberate physical damage is done to residential buildings and schools, coffee shops, and shopping centers. The damage is incurred by firing a few rockets and exploiting the weaknesses and fears of the population on the other side. Above all, the perpetrators exploit the fact that Israel is a westernized, relatively liberal society where the sanctity of all human life is valued, and where media exposure is extensive. Israel cannot respond to a terror attack with terror. We are hardly even able to shoot back at those who fire rockets at civilians.

This is an asymmetric conflict, where the hands of the state are tied by the law, whereas the hands of the other side are free to perpetrate acts that are not bound by any democratic principles or laws. The opponents of Israel often use guerilla warfare and semi-military war tactics; it uses a strategy of disappearance. When it wishes it confronts the army with uniforms and weapons; and when it wishes it sheds its uniforms, enters homes, and brandishes the weapons from a place of hiding so that the military force will expose its Achilles heel.

The task of fighting terror is highly problematic. It involves military activity against perpetrators whose goals are defined as civilian and who hide among the civilian population. The Palestinian population is exploited as a human shield to deter the Israel Defense Forces from preemptive strikes and from rooting out the source of the threat. However, it is the moral duty of the commanders to emphasize the army's practical commitment to the principles of *self defense and the necessity of military force* together with the principle of *restraint in the use of force*. The democratic state is morally committed to protect the lives of its citizens from the danger of terrorism, and that is also a moral stand. Protection should be obtained, for example, by neutralizing the sources of terror and by attaining long-term deterrence – but how far can the Israel Defense Forces go? In a democratic state, the army is committed to adhere to the codes of international law and military ethics. It is committed to develop its officers and face these challenges in a professional way, based on considerations of proportionality, accurate assessment, and discretion. This behavior is expected, even in the face of sharp criticism against the policies of the IDF – especially criticism against the way that the army has planned and carried out missions aimed at neutralizing the centers of Palestinian terror. Nonetheless, there is a need for serious, responsible criticism based on rational considerations and comprehensive understanding of the facts.

The Aim of the Article

The article aims to present the main challenges faced by the Israel Defense Forces in the effort to maintain military ethics in combat, with special emphasis on the types of dilemmas that the army needs to address. An attempt will be made to discuss the learning opportunities inherent in the situations the army has been forced to deal with, and to present a wide range of examples for the development of normative thought and for the professional and ethical development of officers. In the context of topics related to military ethics, the article will deal with the *way* that a just war is waged (*jus in bellum*) rather than with the "just war" per se (*jus ad bellum*) which, in principle, relates mainly to decisions made at the political level.

What is a Dilemma?

The term dilemma relates to a situation in which a choice has to be made between two or more alternatives. However, an *ethical dilemma* is a situation that will often generate an apparent conflict between *moral imperatives*, in which to obey one would result in transgressing another. The dilemma creates a situation of conflict within the individual, and is usually very difficult to resolve. Even though every dilemma reflects internal contradictions between a person's desires, there are various reasons for those contradictions.

- Moral dilemmas: this type of dilemma involves conflicts that derive from motives related to matters of conscience, and to difficulties about what is "right". The dilemma arises when a person has difficulty deciding which alternative is more moral – even if in principle the situation often can be resolved.
- Dilemmas related to costs and benefits: In a "cost and benefit" situation, the dilemma involves different alternatives, where the individual has to decide which alternative will be more beneficial.
- Default: A situation in which none of the available options are chosen, i.e., the choice not to choose.

An ethical dilemma is created when there is a conflict between values, and when a choice has to be made between acts that represent different standards for appropriate behavior. This dilemma represents structural tension between sets of values:

- *Universal social values*: Values that reflect what is right and appropriate in any society. For example, in the postmodern world the prevailing values are based on moral pluralism, which legitimizes differences in perceptions of justice and equity. The situation is further complicated by the fact that a given set of values does not persist over time, and that their importance can change in light of developments in the modern world. Even though all armies have an ethical code, prevailing social values do not remain static. In Israeli society, willingness to give national interests priority over one's personal life is not as high as it was in the past; nor is the perception of military service as a mission as prevalent as it was in the past. Thus, the balance between the values of concern for the needs of the individual versus the needs of the collective has changed over the years – and it is those values that highlight the complexity and problematic nature of decisions regarding various dilemmas.
- *Professional values*: These are the values that professionals set for themselves, irrespective of the organizations that employ them, in order to establish clear professional standards and to protect themselves or those who use their services.

Since its establishment, the IDF has been guided by a code of ethics which includes ten important values.[2] Of those, the three basic values delineated in the Spirit of the IDF are as follows:

---

[2] The ten values delineated in the "Spirit of the IDF" are as follows: Tenacity of purpose in performing missions and drive to victory; responsibility; credibility; personal example; human life; purity of arms; professionalism; discipline; comradeship; and sense of mission.

- *Defense of the state, its citizens and its residents*: The IDF's goal is to defend the existence of the State of Israel, its independence, and the security of the citizens and residents of the state.
- *Love of the homeland and loyalty to the country*: At the core of service in the IDF is the love of the homeland and the commitment and devotion to the State of Israel – a democratic state that serves as a national home for the Jewish People, its citizens, and residents.
- *Human dignity*: The IDF and its soldiers are obligated to protect human dignity. Every human being is of value regardless of his or her origin, religion, nationality, gender, status, or position.

Additional values are:

- *Organizational values*: These are values which derive from the organization's domain of activity, and which the organization has defined for itself in order to attain its goals. The values that guide the behavior of the organization, or influence the professional behavior of the organization's members. They cannot derive from the unique leadership behavior that commanders demand during certain periods and that are consistent with the Spirit of the IDF, such as excellence, modesty, etc.
- *Personal values*: Beliefs, attitudes, and opinions that derive from the individual's internal world, such as religious conscience or values instilled in the parental home. Soldiers serving in the army represent and believe in different values or political and moral perspectives which evidently influence the way they perceive the legitimacy of their own activities or those of the organization.

By nature, war involves a conflict between different values or between different types of moral obligations, e.g., the obligation to refrain from killing and the sanctity of human life versus the right and obligation of self defense or the obligation to save human lives. In the case of a military ethical dilemma, and in the war against terror, the conflict involves a need to address four additional challenges:

- Tenacity of purpose in performing missions
- Preserving the lives of soldiers
- Preserving the lives and integrity of innocent civilians
- Preserving the lives, well-being, and security of citizens of the country

Therefore, when these values conflict with each other, we may ask ourselves several personal and professional questions:

- Do the values delineated in the Spirit of the IDF still provide the basis for solving this problem? In solving the problem, does one value naturally take priority over another?
- Is there a need to adopt new values as a basis for coping with the problem (Kasher & Yadlin, 2005)?
- Is there any value that naturally takes priority over the others, and if so, why? What does that mean? Is it right and fair?
- Would soldiers, the Israeli civilian population, and military commanders see this as the solution to the problem if they are put to the "mirror test"? Would they be satisfied with their decision?
- How will the decision be explained to the international community? Is it consistent with international law?
- Is the decision balanced? Does the alternative allow for a balance between different values or interests?

Learning and Professional Development in Dealing with Dilemmas

The task of dealing with a moral dilemma – beyond being part of the reality imposed on the organization and its staff members – lays the foundations for constant professional development. Because army commanders have to cope with new realities every day, their efforts result in the development of knowledge, understanding, thinking, and experience in the profession – in this case, the military profession. People learn through dilemmas, and the process of deliberation generates new insights about the world: it plays an important role in the process of development.

Professional development is an outcome of coping with situations that pose challenges and stimulate thought; it occurs in formal institutional settings, particularly in military academies or military colleges, as well as "on the job", when the soldier observes senior officers making decisions in their daily lives. In the IDF, formal programs are based on elements of theories, case studies, simulations, meetings with commanders, and self reflection. The dilemmas are constructed as hypothetical cases. In that context, they are carefully formulated to allow for real deliberation about the different considerations, or they are presented as historical and actual case studies and narrated from the perspective of the commanders or instructors.

The reality of combat, however, is a situation that the commander deals with daily in the context of training, and it goes beyond the level of hypothetical cases. In that reality, the commander exercises discretion, examines how his own direct commanders operate, how his superior commanders act, and what the prevailing mood is in the unit, in the media, and among the civilian population. This latent and manifest learning constitutes an important tool for constant exploration of the different levels of the profession, and it is the ethical foundation underlying his activity. However, in formal learning as well as in the field, there is often a conflict between the values that guide the combat operation, and the values that guide other military activities such as humanitarian intervention, maintaining peace, intervention in natural disasters, etc. In many cases, different environments call for different choices.

Individual, group, and organizational learning about issues related to military ethics should not be based on constant examination of gaps. The culture of discourse that focuses on gaps between espoused values and actual values is complicated to implement. In that culture, controversial questions are posed to a group, and emphasis is placed on critical thinking, as well as on analysis of the reasons why we are far from attaining our espoused goals, or why we continue grappling with dilemmas – these are a complicated methods to implement (Raviv, 2005). However, in the hierarchical military culture, learning through a process of questioning is not always consistent with the need to provide clear, well-founded answers and with the need for military discipline, uniform goals, and relations based on authority. An effective discussion of ethical dilemmas in the classroom, in the unit, or on the battlefield will be possible only when the model of authoritative command and compliance with instructions is replaced by dialogue. In that setting, there is often an "institutional truth", but the means of reaching that truth need to be genuine and bold.

Military dilemmas can be discussed along three dimensions, as mentioned above. In the following section, I will demonstrate some of the dilemmas that typify each dimension, as they have found expression in the reality of the Israel Defense Force in recent years.

Ethical Dilemmas – Actions to be Taken in Each of the Three Dimensions in which the Military Operates

When developing ethical awareness, it is important to note the three main entities that the military professional interacts with: 1) the people – the basic body for which he works, i.e., the democratic state and all its citizens; 2) colleagues – commanders, subordinates, the unit; and obviously 3) the enemy. The Spirit of the IDF deals with all three, and adopts the principles and values most appropriate for each interaction.

Dilemmas Relating to Combat against the Enemy

In the introduction to this article, I emphasized the complex dilemma of dealing with terror. However, in the war against terror, the laws of combat should be followed professionally and with restraint. In the war against terror, indiscriminate actions are not taken. On the contrary – attacks are aimed at the sites where rockets are manufactured, launched, and supplied, as well as at armed killers, and sometimes at the leaders who plan, incite, and guide the forces that carry out the attacks. However, as mentioned, the launching sites are situated in dense population centers, on the assumption that they will serve as a civilian shield. For example, thousands of Qassam rockets are fired at localities in the southern region of Israel. Which considerations should guide the IDF in its efforts to cope with those attacks?

An examination of the principles guiding the IDF before a military operation is approved and implemented reveals that the planning processes are extremely careful, even if that aspect is not sufficiently emphasized in the public discourse. Intelligence investigations outline the target of the operation, i.e., to strike at the center of terrorist activity, and provide a picture of the civilian environment in which the proposed operation will take place. Not only does the army have no intention of harming civilians or even causing minimal damage, but the actual objective is to carry out the strike against the terrorists without touching civilians at all. In that respect, the guiding principle is the necessity of military action. That is, military action must be fully justified on the grounds that if it is not carried out, it will not be possible to protect the civilian population of the country. As such, if the IDF attacks a civilian environment, it has to be done as a military necessity (Kasher & Yadlin, 2005). If there is any way of achieving the same degree of success with the same risk to the lives of soldiers and less risk to civilians in the targeted area, then that option should be the preferred one. If that is not possible, then the operation should not be undertaken at all, and efforts should be made to use other strategies. In any case, it is clear that indiscriminate attacks against non-military targets are viewed as unacceptable in the IDF.

However, what happens when the Hamas positions children on the roofs of houses where rockets and weapons are stored? Is it permissible to destroy agricultural fields and homes in an area where Qassam rockets are fired daily? If so, to what extent should this be done, and how is proportionality determined in this case?

Some of the complaints directed against the IDF in the Second Lebanon War related to the army's preference for protecting the lives of civilians on the side of the enemy over protecting the lives of Israeli citizens. It has also been argued that the IDF could have made more of an effort to stop the rocket attacks on the civilian population of the north and reduce the number of Israeli fatalities. In any case, when innocent lives are taken inadvertently, the sword is turned in the other direction. Unfortunately, in the struggle for public awareness, it is the picture of innocent civilians falling victim to bloody acts that makes the ultimate impression. In the struggle for public awareness, Israel does not have the advantage. Israel has not succeeded in explaining that civilians are killed in air strikes because the homes under attack are actually factories that manufacture Qassam rockets. In the international media, Israel is unable to give momentum to the fact that in February 2008, when the Hamas shelled the *Sufa* crossing, 60 trucks entered with food from Israel as part of an effort to provide humanitarian aid. This message also has important implications for the morale of Israeli soldiers. Oftentimes, the "struggle for public awareness" is not affected by a given military operation. Rather, it is affected by a picture in the paper.

Finally, it is important to mention that in cases where accepted values are violated, or in cases where normative patterns of behavior should have been maintained but criminal acts were committed despite serious deliberations and thought, the necessary disciplinary measures should be taken to root out the failure. In cases where deviant command culture and norms are identified, they should be severely punished as an example for all to see. This kind of treatment relates to training, command, and practice in the field.

Dilemmas of the Army vis-à-vis the Political Entity it Serves

It often happens that soldiers and commanders don't identify wholeheartedly with the tasks they are given – and army service is often based on a political social ideal such as Zionism. The soldier and commander can voice reservations about the moral legitimacy of going to war or the legitimacy of what is done in the context of war. Even if the military mission is professional, legal, and binding, personal identification with the mission lends validity and meaning to its implementation. What happens when there is a complex discourse between a soldier and commander regarding the military order that the soldier was required to carry out – even if the order was totally legal? What dilemmas can the order arouse, and what is their significance? A case in point is the implementation of the Disengagement Plan in the summer of 2005. When the disengagement from the Gaza Strip took place, 25 settlements in the Gaza Strip and the Northern West Bank were evacuated. This was a dramatic test for Israeli society and democracy, and posed a complex challenge to Israel's interagency security cooperation.

Even though the disengagement was the outcome of a decision made by the government of a democratic state, it had the potential to cause a serious rift in the army. Some of the soldiers who were given the command to implement the disengagement were religious people who viewed the settlement of the land of

Israel as a sacred duty and obligation. As such, the evacuation of civilians from their homes aroused intense, complex, and harsh emotions. Thus, they grappled with the question of identity and had a conflict between loyalty to their rabbis versus compliance with their commanders. The disengagement was implemented against the background of a deep-rooted and long-standing disagreement about the future of the "territories", and the act of disengagement was perceived as a declaration of a permanent border of the state of Israel. The army dealt with that process by distinguishing between two concepts: "justness of the ideological cause" and "justness of the military cause"[3] The idea was to maintain an apolitical position and continue implementing government policies as a type of professional stand, and as an expression of the concept of representation and loyalty. This was done through just use of force, restraint, proportionality, military solidarity, sensitivity, and respect for the evacuees. Conceptual organization of the dilemma and development of relevant knowledge relating to military activity in an era of controversy does not solve the dilemma. However, it can be a guideline for soldiers and their units when they are performing a task in the present or future.

Dilemmas Concerning Relationships between Soldiers and their Comrades

Given the nature of its activity, the military organization has to adhere to basic values of standing by its units and commanders, where every effort is made to successfully complete the missions of the unit. Comradeship in arms, willingness to help others and put oneself at risk, and even willingness to sacrifice one's life for others are all basic values in any army. In the Second Lebanon War, there arose a dilemma which posed a conflict between tenacity of purpose in performing missions and the drive to victory on the one hand and the values of comradeship and personal example on the other. According to the Winograd Commission Report (2008), the fulfillment and maintenance of some of the values of the IDF during the Second Lebanon War was deficient. These deficiencies were not because the norms of operation themselves were bad or inappropriate. Rather, the norms were not properly maintained. The Commission placed special emphasis on norms relating to tenacity of purpose in performing the mission and the drive for victory even in the face of difficulties and casualties.

The Winograd Commission found that some people at higher levels of command gave orders to stop advancing or to stop fighting rather than let the rescue and medical forces evacuate wounded soldiers. It appears that these high levels of command exhibit excessive tolerance for constant delays, which prevent forces from performing their mission. In addition, commanders tend to approve

---

3   Concepts coined by Prof. Asa Kasher.

or delay missions on the grounds that the forces are not completely ready to act, or on the grounds that weather conditions are not optimal. The fighters and commanders have not always understood why it is essential to sacrifice themselves and take risks in order to enter an area that will be evacuated immediately after the battle. The message regarding the importance of minimizing casualties, which was conveyed at the higher ranks of the military command and at the political level, affected the planning of the mission and reinforced the ethical orientation dictated by the value of comradeship – as reflected in the need to care for the wounded.

In addition, the Commission Report mentioned weaknesses in implementing the value of personal example. When the battles were taking place, some of the commanders did not position themselves alongside their combat forces; they stayed in war rooms instead and maintained control from there. The value of personal example is perceived as sharply contradicting the desire to enhance the capacity to command from a distance using advanced technology. It is difficult to convey a message to fighters that emphasizes the importance of combat and of performing missions when the commanders are sitting on the sidelines. The importance and necessity of war are not self evident, and in light of the conclusions of the Winograd Commission, it appears that over the years, the practice of commanders staying with their forces in the field has been eroded.

Future Dilemmas and Implementation of Solutions – Summary

The presentation of various dilemmas on the three areas in which the army operates highlights the accuracy of the statement that "reality is the playing field"; reality is the richest, surprising, and most complex context on earth. It affects the military career in several ways – areas of knowledge, tools, and the values by which it operates. The more changes occur with regard to the nature of combat, the orientation toward individualism, and the widening of social gaps, the more complex techniques will have to be employed in dealing professionally with the dilemmas.

The military institutions are responsible for identifying the dilemmas in an organized way, but the commanders also bear responsibility in the context of encounters and activities in the field. In the endeavor of putting values into effect it is no less important to engage in constant identification of potential gaps, and in deliberating possible reasons for the existence of those gaps.

Even if it were possible to solve the dilemmas in certain cases, it is necessary to consider how the values and principles we choose can be instilled, what measures and signs indicate that the values and principles have been instilled, what terms and concepts should be used to indicate those values, and what organizational measures need to be taken to promote them. How can comprehensive, extensive strategies be constructed to deal with deviations and failures – if those are found? And how can those issues be addressed in training courses? What forums should be used in order to talk about those issues and to effectively convey the messages about what we are not prepared to compromise on? Of course, when it comes to solving dilemmas it is important not only to talk, but also to make the words consistent with actions. In that context, it is important to determine appropriate rewards and sanctions, and to constantly seek practical mechanisms of assimilating them into the military organizations – in routine situations and in situations of emergency.

Instilling values is the greatest challenge faced by armies. Ethical dilemmas are *part* of the act of combat, and are not a separate area that should be ignored. In the future, officers will have to brainstorm in order to come up with ideas about how to fight terror and ideological fanaticism. They will have to have a good understanding of the political aspects of war, the complexity of military-social relations, the operational implications, and the gaps in consensus about some applications of international law – for not everything that is legal is also ethical, and vice versa. When professional standards are raised, values are enhanced. That is how the message is conveyed to society that the army understands the magnitude of its mission and the extent of its responsibility. Only thus will armies – including the IDF – be able to continue enjoying support from the public, which is so essential.

Bibliography

Kasher, A. & Yadlin, A. (2005). Military ethics of fighting terror: An Israel perspective. *Journal of Military Ethics, 4*, 3-32.

Raviv, A. (2005). Teaching values for military professionals. In E. Micewski & H. Annen (Eds.), *Military ethics in professional military education revisited.* Frankfurt am Main: Peter Lang.

Winograd Commission (2008, January). *Final Report, Chapter 11: Summary and Conclusions on the IDF.* Jerusalem: Author.

Amira Raviv

Developing Senior Leaders: Challenges, Methodologies, and Dilemmas

Introduction

The abundance of literature on leadership emphasizes its relevance. The topic of leadership has continued to concern researchers for centuries, and the literature attests to its changing, complex nature. In recent years, there has been a need for further distinctions between different levels in the hierarchy of leadership. In that connection, several questions arise: How is senior leadership different from junior leadership? Are the teaching methods used to educate senior officers different from those used to train younger officers? What professional perspectives affect the development of senior leaders? And what methodological difficulties and challenges are encountered by military colleges in the process of developing senior leaders?

The Task Environment of Senior Commanders: The Changing Face of Warfare

Senior commanders operate in an environment that has become increasingly complex. Today, wars continue for a relatively long time, and they are similar to cultural wars of attrition which are directed to the arena of national consciousness. Giora Eiland (2007) analyzed six dimensions of change in the nature of warfare that have affected the environment in which senior commanders operate:

First, there has been a change in analysis and adaptation of variables in asymmetric wars that take place in densely populated civilian areas. That issue has posed three challenges. One challenge relates to intelligence – not only determining *where* the enemy is, but *who* the enemy is. Another challenge relates to identifying an enemy, when the enemy may alter in light of rapid changes in diplomatic relationships and political definitions. The third challenge relates to characterization of a relevant target. The enemy is not a standing target, nor does he stay in one place for long. Rather, he is constantly moving, and his location can change within a matter of minutes and even seconds.

Second, there has been a change in the relationships between the political and military systems. In traditional wars, the political system notifies the military that it is expected to win the war, and the politicians deal with its outcomes. The definition of victory is usually clear, and can be interpreted in military terms. It is demarcated in terms of its area, time frame, and political framework. In a low intensity conflict, the boundaries between the political and military systems are sometimes blurred. It is not always clear what goals should be achieved, and strategic objectives are not always properly defined. In this situation, there is a need for a different type of dialogue that is not based on hierarchy. It is important to discuss not only how to attack, but to question whether it is right to attack. Toward that end, frequent discussions are held between military and political officials. Every political action has implications for security, and vice-versa. Officials on both sides need to meet more frequently in order to discuss the current state of reality. In addition, there is a growing connection between opportunism in the political leadership and public loss of faith in the government, its leaders and its institutions, as well as a growing sense that political leaders are sometimes involved in corrupt decision-making processes (Ben-Yishai, 2007). In light of these developments, there is an increasing need to create a new leadership in the field of national security. Those leaders should be committed to moral values, and should emphasize the responsibilities of senior officers in the defense establishment – especially military commanders.

Third, implementing organizational and procedural change, with emphasis on achieving jointness. Current wars are no longer characterized by a clear definition of total peace or all-out combat. There are countless situations that require a different type of division of authority and coordination between various entities. For example, an interagency dialogue was held in the United States on the topic of Homeland Security after the Twin Towers attack in 2001. Similarly, an alternative type of dialogue was required in Israel to fight terror. In that process, the leading security agents (the Mossad, the Military Intelligence Directorate, and the General Security Services) were called on to establish new mutual domains of responsibility. This was important, because some of the coordination in these domains transcends institutional boundaries. Clearly, the responsibility for establishing this kind of dialogue lies with the senior officials in the national security system.

Fourth, coping with the challenges of new information technology. More advanced, destructive technologies are not always effective. In fact, sometimes the opposite is the case: technologies need to be adapted to the new threat, where the goal is not necessarily to achieve maximal destruction. On the contrary, there is need to use less lethal technologies such as small unmanned aerial vehicles.

Fifth, proper and wise use of the media. The senior commander is influential and manages an environment in which there is intensive media coverage. In that context, it is important to promote a dialogue with civilians – especially in order to gain national and international legitimacy for military operations. That kind of dialogue will allow for freedom of action and sometimes enable the military operation to be prolonged if necessary. It is critical for the senior officer to be familiar with the media and competent in conveying and understanding messages through the media.

Sixth, debunking myths. In the context of modern wars, some myths are shattered. One myth relates to definitions of the duration of war. For example, Major General Eiland notes: "If we defeated four armies in six days in 1967, then how many days do we need to defeat only a few thousand Hezbollah fighters?" (p. 17). Another myth relates to the number of casualties in war. Some confrontations are "wars of choice". In those contexts, there is an unrealistic expectation that our side will win the war without endangering our soldiers, because we have sophisticated weapons that can attack from a distance. A third myth relates to the ability to avoid harming innocent civilians. We will support an all-out war with the Hamas as long as the casualties are solely enemy soldiers. But when the television shows horrific pictures of killed children, we begin to have profound doubts and reservations. In those cases, we criticize ourselves and ask disconcerting questions about the justification of military action. A typical question raised in Israel is: "What happened to us?". It's not always clear why we're here and what we're fighting for; whether the price is worthwhile. A fourth myth relates to the ability to achieve a decisive victory. We are prepared to pay the price of war if we achieve a decisive, clear victory where the enemy surrenders or gives up the will to fight. However, in the new war the perception of victory depends on one's world view, and is based on a narrative that is difficult to change.

In light of that situation, senior officers are called on to win the new types of wars that characterize the 21st century. They need to prepare for different types of conflicts by showing flexibility and versatility in the use of security systems and recognize limitations in the use of military force. In the effort to gain enhanced legitimacy from society, they need to employ a strategy that recognizes the possibility that there will be violent confrontations and that people will be killed. They also need to be attentive to critical voices which question the justification for their approach and the purity of arms in every instance. In addition, they need to be able to explain the actions of the military and gain support through a dialogue with citizens and soldiers from the entire political spectrum of society.

The Characteristics of Senior Officers

In light of the changing security situation, it appears that the characteristics of senior commanders – some of which have been characterized and defined in numerous studies (e.g., Altman, 1999) are more relevant now than ever.

Maintaining a growing physical distance in relationships with subordinates. Due to the distance, which intensifies psychological projection and attribution, the senior commander impacts his subordinates through symbolic leadership. Units are split into sub-systems, and rarely operate as an organic whole. In this context, the commander's challenge lies in enabling his soldiers to simulate an organic and yet individual framework and maintaining an atmosphere of learning and cohesion (Kaplinsky, 2007).

Attaining desired outcomes through other managerial levels. Because the leader operates through intermediaries and intermediate level staff, he needs to be able to develop systems to influence his subordinates. In that context, he needs to choose – or at least be involved in choosing – managers who will know how to convey his messages to the lower levels of the organizational hierarchy.

Understanding the limitations of control. The senior officer is responsible for handling a tremendous volume of information and knowledge. In an environment where the mission involves complex tasks characterized by feedback, dependence, and unclear cause-effect relationships, the senior officer grapples with abstract ideas based on open questions, concepts, and symbols.

Establishing "jointness". Senior officers need to work with social networks and build partnerships based on relationships between different units and organizations. In light of this situation, there are several characteristics that are essential for working in those environments and dealing with those challenges. Above all, senior officers need to develop abilities for strategic thinking and coping with ambiguity. They need to think in terms of complementary opposites, taking into account a complex network of causes and effects and the emergence of processes over time. They also need to identify patterns that will enable them to understand and operate in a complex arena. In that context, it is essential to perform multidimensional tasks which might even be contradictory and inconsistent (e.g., there may be conflicts between political, legal, military, and economic perspectives). Senior officers also need to know how to operate in unfamiliar situations without clear instructions and how to make rapid transitions between different agencies and functions in a constant process of integration and learning. At the level of learning, senior officers have to be able to interpret situations in terms of regional and global trends. Toward this end, it is necessary to engage in strategic and innovative thinking, to construct new paradigms of reality, and to break into a given sphere of influence. In addition, constructing new conceptual frameworks, constantly introducing innovative ideas (Ben-Ishai, 2007), developing analytical tools, and formulating organizational goals in a dynamic world is also needed. Commanders need to be able to manage complex emotional situations, to cope with ambiguity and isolation, and to bear the heavy personal burden of managing those processes. They also need to manage and influence multiple networks. This includes working in a political environment, as well as developing collaborations, strategic agreements, and alliances. In those contexts, the commanders need to have sensors to detect hidden agendas and underlying inter-organizational developments. In addition, they need to be able to manage teams of experts which specialize in areas where they lack specific knowledge.

Another interesting dimension that has emerged in recent years is the ability to *communicate in multiple languages* – to convey messages and influence people through speeches and statements, using tactical, systemic, political, strategic, media, and civilian semantics that are adapted to the target audience. Most importantly, the commander must have a high level of self-awareness, understanding the need to clarify himself and recognizing his personal style and his own strengths and weaknesses. In that context, he should be able to identify hidden assumptions, know to ask difficult questions, show modesty, and recognize the limitations of his own knowledge. Finally, the commander should know how to develop a source of internal authority.

Values Unique to Senior Leadership

In light of the above, one of the essential core characteristics of the senior officer's role is his involvement in shaping an ethical operating environment in the military. In that context, one of the officer's major concerns is to maintain the moral image of the army. Because morals and values are a fundamental part of combat, anti-terrorism warfare generates a conflict and arouses complex dilemmas on several levels: how to protect the lives of innocent civilians and restrain power when necessary; how to preserve soldiers' lives and still protect one's civilian population; and, no less important, how to establish tenacity of purpose in performing the mission.

Although all of the values of the armed forces are relevant to any combatant, it seems that three main values are most characteristic of the activity of senior leaders:

Responsibility: Responsibility means contributing to events, processes, methods, and results. In contrast to younger ranks, it is especially important for senior commanders to assume personal responsibility for outcomes, responsibility for promoting and implementing ideas, and for the people and agencies engaged in the mission. Senior commanders assume responsibility for maintaining relevance and innovation in learning, as well as for engaging in a dialogue with agencies that collaborate with the national security system. In addition, the senior commanders represent the military vis-à-vis the civilian system – even when they personally disagree with the decisions made at the political level. Perhaps most important are the leaders' value priorities for operating in complex, ambiguous environments.

Personal example: Setting a personal example is an important value for any kind of leadership. Senior officers, more than soldiers or commanders at any other level, should serve as a role model of collegiality, honesty, loyalty, and moral integrity in decision-making. They should be role models for constructing and operating the military force. The role model is expressed through serving as a personal example at the level of the smallest things. Due to the impact of symbolic leadership, this value is enhanced even more.

Professionalism: Professional practice involves the following elements (Kasher, 2005): a systematic body of relevant knowledge; systematic proficiency in solving relevant problems; constant improvement of relevant knowledge; local understanding of the claims of knowledge and methods; and global understanding of the nature of the system of knowledge and proficiency (ethics). Not only is the senior commander responsible for promoting all of these aspects, but he is also required to master a new realm of knowledge for the sake of the operational and strategic mission. In that context, the officer aspires to achieve excellence and to broaden his local and global understanding. The leader must learn to develop a professional language appropriate for mediating between the strategic and tactical levels. However, he must be careful not to create misunderstandings, and he must make an effort to initiate a dialogue without upsetting the hierarchy of command.

Senior Leaders and the Learning Process: Characteristics and Challenges

The Paradox

Even though it is clear that such complex behavior requires formal education and learning, it is paradoxical that the higher a commander's level of seniority is, the more barriers he faces in the learning process. Meaningful learning poses challenges, as the individual confronts gaps or lacunae in his own knowledge. Therefore, senior leaders often have difficulty assuming the role of students – perhaps because they leave a situation in which they are in control and unconsciously realize that something is lacking. Most of the knowledge is not acquired on premises that are marked by a sign "Studies in Progress". In general, learning does not take place in the courses per se. Rather, the courses facilitate the learning process. For most senior officers, learning takes place "between the lines", and not always in the formal setting of programs at military colleges. In addition, meaningful learning does not depend on what the lecturer or instructor says. Rather, it depends on the significance that the participants in courses attribute to the knowledge. Sometimes spontaneous learning events are perceived as more critical, interesting, and memorable than experiences that are anticipated and planned. However, because of role responsibilities, loneliness, and emotional isolation that commanders encounter in their role, it is important to establish planned, formal educational settings on the one hand, while giving the participants in those settings room for creativity and exploration on the other. Therefore, it is important to consider the needs of senior leaders as part of a world view that guides the development of curricula for that population.

Learning Attributes of Senior Commanders

Senior commanders have several attributes that are drastically different from those of junior officers (based on the theories of Knowles, Holton & Swanson, 1998):

1. They have *extensive life experience* which is organized into existing schemata and is part of their self-identity. As learners, they seek to identify the connections between the learned material and their past experience, and they expect their experience to be recognized. Failure to acknowledge their experience is tantamount to disregarding their identity.
2. Senior commanders have a *sense of psychological seniority and maturity*, and they prefer to decide their own future path, destiny, and activities. They have a deep psychological need for independence and do things that are consistent with their status as adults.
3. Senior commanders have a *practical, pragmatic perspective*. They look for immediate relevance in the learning process and gauge its contribution to their professional and personal development. They need to know why they are learning a specific topic and what use it is to them.
4. They have intrinsic motivation and are driven toward self-fulfillment. As such, they are influenced less by external factors. Adult learning theory assumes that the basis for learning is a functional need. However, differences in needs are varied: There are those who learn for the sake of future roles, those who seek to achieve a certain goal, or simply out of love of knowledge. Pedagogic

emphasis should be placed on intrinsic motivations such as self fulfillment and self-esteem, in contrast to extrinsic factors such as sanctions.
5. As autonomous, critical adults, they have a *skeptical attitude* toward ideas presented to them in the learning process. Therefore, learning should be based as much as possible on dialectical, critical thinking. In addition, management of educational programs for senior commanders should focus on creating a learning environment characterized by respect, mutuality, and pleasantness. Those aspects are the foundations of situations and learning environments appropriate for adults.
6. Senior commanders learn best through *activity, involvement, and solution of practical problems* relating to their organization and their jobs. They learn best when the material has personal significance for them and when they are presented with challenges.
7. Senior commanders have a well-developed sense of *"wasting time"*. They have a low stimulus threshold when it comes to activities they perceive as too time-consuming or inefficient.
8. The adult student himself is a *source of knowledge in a group*, in addition to the teacher. Therefore, peer learning is an effective learning technique for adults.
9. *Egalitarianism and fostering egalitarian relationships* between the instructors and students is a basic principle of adult education. In that context, the instructors encourage students and involve them in setting the goals, content, and methods of learning, as well as in evaluating the learning process and its outcomes. The curriculum of the course is designed according to the needs and interests of all groups. In addition, the participants are invited to engage and take part in facilitating the learning process itself.
10. It is assumed that the group of participants in the course will be *heterogeneous* in terms of their job positions, personal styles, age, organizational tracks, and geographic background. Therefore, they should be given opportunities for social interaction in the learning process, they should work in small groups, and they should be given space to choose the pedagogic and thematic tracks and issues that interest them.

Programs for Development of Senior Leaders: Methodology and Main perspectives

In light of the above, a curriculum designed for senior officers should be based on a range of considerations which encourage the participants to learn in an environment that emphasizes three principles (Raviv, 2003):

Maximal involvement and active learning: Giving senior officers an incentive to assume responsibility and control, to choose and influence the learning process, and to bring some of their own material to discuss in class. Thus, learning is not based on passive processes where the curricular content cannot be challenged. Rather, it is based on analysis and interpretation of material, with the participant involved in setting his goals of the educational process, and learning is independent and exploratory. In that context, the program facilitators are also responsible for arranging frequent meetings between the Chief of Staff, the senior board, and the participants, as well as for inviting the learners to participate in planning the agenda of the meetings.

Utilizing a variety of methods in the program: In light of the broad range of needs and expectations of the participants, and considering that they have a wide array of interests, it is important to integrate theory and practice in the program. In addition, the training program should incorporate experiences and models, group work, reading material, case studies, personal self inquiry questionnaires, meetings with senior officers, and meetings with inspirational political leaders. All of these methods combine a *dialogical environment* and constant interaction between theory and traditional knowledge on the one hand, with skepticism, practice, and conceptualization on the other.

Promoting a critical, creative study atmosphere. The program needs to let the "stormy winds" of the environment blow into the classroom and open the door to new perspectives, even if the atmosphere in the classroom might become turbulent at some points. The participants have an opportunity to express meaningful ideas that undermine existing paradigms. Nonetheless, the facilitators are called on to maintain an atmosphere of tolerance, which encourages and respects processes of change and extending of the "comfort zone". Additionally, the program should emphasize ethical dilemmas, with the understanding that even leaders in high positions don't always have clear answers to every situation.

These curricula often provide a genuine opportunity for organizations to develop valuable knowledge. That kind of learning, which occurs away from the battlefield and daily problems, invites participants to concentrate and to experience "being" rather than just "doing". In this pedagogical environment, latent knowledge emerges into articulated knowledge. If the programs are properly managed, they allow space to examine phenomena that seem marginal or yet go unnoticed in the organization. These research activities and "think tanks" have the potential to be transformed into practical models that can be utilized in the country's defense system. In national defense colleges, special care is given to new interagency models that are developed from non-sectored learning situations. However, this organizational knowledge is only one important part of learning. The most profound pivotal event that has to take place in the program is the development of an individual sense of self efficacy and understanding, in addition to moving the participants into a position of influence, change, and self awareness.

Classic Methods of Teaching in Courses for Senior Leaders

There are many teaching methods that can be valuable in different kinds of learning situations. However, some have proven to be more appropriate or more effective for the population of senior leaders. The following are a few examples:

- Leadership and Management Workshops in Small Groups
  - Workshops that combine personal diagnostic questionnaires for leaders and 360° feedback systems combined with personal coaching. The challenge here is to offer added value to what the officer has learned on his own through his past military career.
  - Workshops that allow participants to discover personal values or challenge existing, lacking, or partially hidden values.
  - Consultation peer groups that deal with personal dilemmas related to leadership: members of these groups discuss dilemmas with each other and offer advice and solutions based on their own experiences.
  - Experiential workshops in nature, i.e., in field conditions. In those workshops, the participants can experience and conceptualize leadership and team work in a setting that is fun and challenging.
  - Role analysis workshops which enable exploration of relationships between "the self", "the role", and "the organization". This is a strong tool enabling and allowing participants to explore personal fixations and potential regressive patterns that dominate them from early states.
  - Business theaters enable analysis of real-life situations in management. This kind of workshop is conducive to implicational learning and simulations of work situations through role playing. In those contexts, situations and responses are analyzed through the medium of drama.
  - Explorational dialogue on leadership phenomena.
- Lectures
  - Renowned academic experts presenting strategic, organizational, and social models.
  - Key speakers and business leaders lecturing on decision making topics and parallel management challenges in similar defense institutions or business sectors.

- Directors from the organization – at senior levels or retired – presenting a role model for senior leadership, with emphasis on managing crisis situations or managing change.
  - Holding panels where participants present divergent and conflicting perspectives on dealing with a problem.
- Case Studies
  - Analyses of complex phenomena in the context of the real world enable experiential learning based on the life of the organization in laboratory conditions. The analysis touches on issues on the emotional, cognitive, and behavioral levels, using systemic, interdisciplinary thinking. This method prepares the participants for a world that requires critical thinking and formulation of persuasive arguments, in a limited time frame and with incomplete information (Raviv, 2008).
- Simulation
  - Creating reduced and abstract schemata of reality which reflect complex phenomena. The simulations enable participants to experience policy-making, planning, and decision-making, in addition to reconstructing critical and strategic events while developing alternatives that are not revealed in other circumstances. The method challenges latent assumptions and attitudes, enabling participants to experience change and gain new insights (Pulkka & Raviv, 2007).
- Coaching and Personal Counseling
  - Choosing development-oriented counseling in a protected situation where the counselor is a personal resource for deliberating and testing reality.
- Mentoring from Senior Leaders
  - Indirect instruction from a senior leader who is not the direct supervisor, e.g., a retired senior leader in the organization, serving as a model for learning and working, and who has extensive experience relevant to coping with reality.
- Meetings with other Audiences – Benchmarking
  - Meetings with senior leaders from parallel or different organizations, aimed at enhancing the participants' repertoire of insights about other organizational cultures, with an emphasis on case studies.
- Tours
  - Emphasis on the level of values and identity: Tours that bring senior leaders closer to the roots of their national, social, and religious identity and to the sources of their faith. The tours enhance the robustness of values and deepen the leader's knowledge of the country, its population, its social backbone, and the social fabric at the basis of the leader's unique identity and sense of purpose.

- Independent Self directed Learning
  - Curriculum design that puts formal emphasis on self directed learning and allows time for independent learning, reading texts, and interpretation of texts as a mutual process within the learning groups.
  - Inviting the participants to teach the parts of the curricular units, and to gain a monopoly on knowledge in those topics as true experts.
- Open Space
  - Opportunities to explore and spearhead topics that are important to senior leaders, to their future, and to the future of the organization. This is a technology that allows for spontaneity and independent organization from a position of passion and influence.

Of course, not all of these methodologies can be introduced into one curriculum. However, multiple teaching methods create that synergy of effective development. Those methods should be constructed as part of a coherent rationale that is consistent with current needs and recent events, as well as with the characteristics of the unique group and the goals of training.

Some Concluding Thoughts Regarding the Challenges Faced by the IDF

Notwithstanding the processes discussed above, there is a dispute on the question of whether it is possible to improve moral conduct through teaching (Talerud, 2007). It also seems that the challenges of teaching senior officers are so great today, that it is nearly impossible to succeed in accomplishing that pedagogical mission. Furthermore, although many of the suggestions raised in the article have already been implemented in the curriculum for senior officers in the IDF, we are still far from achieving the objectives.

Can we learn something from the experience of the IDF in that mission? The IDF is at the core of public discourse as a people's army. In that context, issues related to the morality of combat have been raised, such as the questions that were posed after the Operation Cast Lead (the 2008 Gaza Campaign): Did the commanders demonstrate moral discretion? Did they preserve the sanctity of life? Is the IDF still "the most moral army in the world", as it claims? In contrast to the biased image that has been portrayed in the media, senior commanders in the IDF give top priority to investigating the truth and analyzing the operations that were carries out. It is well known that most of the combat took place in densely populated residential areas. Although the Palestinian side sustained hundreds of casualties, the senior commanders who led the mission made concerted efforts to preserve the rights of civilians in the war zone. The operation was initiated as a response to tens of thousands of rockets that were deliberately positioned behind a human shield of civilians and fired for months at the heart of Israeli cities. The IDF took extraordinary measures to inform the civilians in Gaza about the targeted areas in order to enable them to escape, in addition to allowing huge amounts of humanitarian aid to be brought into Gaza. Israel had no choice but to defend itself. However, mistakes in war do happen, and should be dealt with through profound debriefing and other measures. At the same time, commanders need to disclose phenomena that violate norms and deal harshly if found, whereas phenomena such as heroic acts need to be reinforced and strengthened. In that process, mistakes, deviations, and ethical dilemmas should be incorporated in the curriculum in subsequent years as case studies which the soldiers can learn from. It should also be noted that senior officers in the IDF face problems within their own forces. Recently, for example, religious Jewish soldiers have threatened to disobey orders if they are commanded to evacuate Jewish settlements or illegal outposts in Northern Samaria. The brigade commanders should deal harshly with any disobedience and toxic political debate running the risk of penetrating the professional military world.

These are just some of the complex challenges that IDF commanders have to contend with. If they themselves stray or make morally biased decisions, they are treated with extreme severity and are forced to leave the service. That is because high ranking commanders play a vital role in sustaining the moral backbone of the army and the nation itself.

Commanders in the IDF operate in a conflictual reality, where they are confronted with questions at home and outside, and where they grapple with the task of clarifying the boundaries between what is allowed and what is prohibited, as well as with clarifying norms in their task environment. In light of the differences between individuals in terms of values, there must be a greater level of clarity in commands in the junior ranks of the military hierarchy. Moreover, the presence of commanders in the battlefield is critical, because they are the ones who point the younger soldiers in the right direction and because they instill values in them by providing a role model for them.

In sum, senior commanders have to develop uncompromising confidence in the justice of their professional actions and the *raison d'être* of the military and state in which they operate. Will military colleges succeed in developing such an excellent senior navigator who is able to lead the soldiers effectively through the stormy oceans they face? This question has yet to be answered.

Bibliography

Altman, A.(1999). Senior Leadership. In I. Gonen & E. Zakay (Eds.), *Leadership and Leadership Development – from Theory to Practice* (pp. 337-368). Tel Aviv: Ministry of Defense Press.

Ben-Yishay, O.(2007). From Leadership to Senior Leadership: How Do We Educate the Senior Civil Servants in Israel? In P. Yechezkely (Ed.), *Military Leadership. The "Broadcasting University" Series* (pp. 113-136). Tel Aviv: Ministry of Defense Press.

Eiland, G. (2007). The Changing Nature of War. *Strategic Assessment, 10* (1), 13-18.

Kaplinski, M. (2007). The Low Intensity Leadership Challenges. In P. Yechezkely (Ed.), *Military Leadership. The "Broadcasting University" Series* (pp. 40-50). Tel Aviv: Ministry of Defense Press.

Kasher, A. (2003). Professional Ethics. In G. Schefler, Y. Achmon & G. Weil (Eds.), *Ethics in psychology* (pp. 15-29). Jerusalem: Hebrew University Magnes Press.

Knowles, M., Holton, E. & Swanson, R. (1998). *The Adult Learner: The Definite Classic in Adult Education and Human Resource Development.* Texas: Gulf Publishing.

Pulkka, A. & Raviv, A. (2007). *Military Ethics Instruction – the Educational Challenge of the Case Study Method. Tiede Ja Ase, 65*, 345-362.

Raviv, A. (2005). Teaching Values for Military Professionals. In E. Micewski & H. Annen (Eds.), *Studies for Military Pedagogy, Military Science & Security Policy. Vol. 9* (pp. 55-59). Frankfurt am Main: Peter Lang.

Talerud, B. (2007). Ethos in and Ethics for Today's Armed Forces in the Western Societies. Why Bother about Military and Ethos? – Setting the Stage. In J. Toiskallio & H. Annen (Ed.), *Ethical Education in the Military: What, How and Why in the 21st Century? ACIE Publication* (pp. 63-84). Helsinki: National Defence University.

Antti-Tuomas Pulkka & Amira Raviv

Military Ethics Instruction – The Educational Challenge of the Case-Study Method[1]

Introduction

As Micewski & Annen (2005) put it, the issue of ethics in military education from the military pedagogical point of view is not only a question of ethically appropriate behavior but rather the challenge of actually teaching this to the trainees. Education, here, should be understood in an extensive framework that is profoundly culturally constructed and therefore deeply dependent on traditions and inherited manners and knowledge; it is not only management of knowledge to be conducted effectively and in measurable ways (e.g., Micewski, 2005; Toiskallio, 2007; Tomasello, 1999; van Baarda & Verweij, 2006). Fleck (2005) quite clearly sums this difference up by stating that although people have usually developed the necessary qualities to act in an ethically sound manner they still may and do act against their better judgment.

Kasher (2003) views the inquiry by a profession into its ethics as an inquiry into ethics itself, since professionals will understand better than others what should and what should not be done within the profession. Ethical principles are not sublime ideals, but rather rules which define desirable activities of people within their professions. Ethics as an assemblage of values and norms directs the professional as to how he should act within his profession. Consequently, ethics is in a sense equivalent to professionalism. Discussion on ethical issues in the profession constitutes a progressive step within it, bringing about an improvement in the skills of normative thinking.

This study will focus on the aspect of methodology of teaching ethics as one possible explanation for the qualitative differences in the learning of ethics. Effective and educationally sound teaching that aims at deep understanding cannot be based on methodology defined by practical economy and action-control beliefs only. The research question of this study is quite plainly put forward by Talerud (2007) as he asks, "[i]s it possible to improve people's moral conduct through teaching?" or as Robinson (2007) argues, "it is one thing to say that soldiers

---

[1] The views, opinions, and interpretations are the authors', and they do not necessarily represent the official national views or those of the authors' parental organisations. This paper has been previously published in 2007 under the Finnish title "Sotilasetiikan opetus – tapaustutkimuksen metodin haste" in *Tiede ja Ase, 65,* 345-362.

will have to undergo ethics training, it is quite another to ensure that they learn the right lessons". Ethical competence is something universally desirable, and nations generally agree on this. It is, however, the definition between the right and the wrong that is culturally related and disputed (Fleck, 2005; Heinonen, 2002). To shed light on this profound matter, we wish to discuss the basis for some techniques in teaching and how the results of this work should assessed.

We present considerations on how to formulate tools for the teaching of ethical matters in a framework constructed on theoretical themes of active learning and self-regulation in learning. The nature and definition of ethics are discussed to form a basis to elaborate arguments on instructional implications. The discussion is followed by a short comparison of two samples of pedagogical materials (Dutch and Canadian) and furthermore by a description of the IDF case study-method. As a conclusion, we wish to lay a foundation for further practical research on ethical education in the military.

Review of some observations on the nature of the ethics

The nature of war and conflict as well as the common sense of military necessities are usually brought forward to justify actions and their direct or collateral consequences (e.g., Fry, 2006). It is not a secret that war, as General Marshall has been quoted to have said, is hell. But war in itself has no nature as such, it is the ultimately horrible human tragedy that should be avoided at all costs, and if it cannot be avoided, it is to be fought with minimal human suffering (e.g., Reichberg, Syse & Begby, 2006). No explanation is tolerable if it refers to something metaphysical or a divine entity or being that lures combatants to act inevitably in one way or another. Decisions are made by human beings, and therefore they are only as good as their makers' imagination. Imagination is always culturally dependable and therefore the actions of armed forces in conflict are oriented by culturally elaborated guidelines (Fry, 2005; Tomasello, 1999). The dilemma of ethics is becoming more and more troublesome in the ever more complicated environment of security, as the earlier guarantees of peace and security are not enough or no longer reasonable (Heinonen, 2002).

Heinonen (ibid.) elaborates on the idea of *global ethics* and delivers two of its assumptions. First, global ethics concern the basic values and other profound premises common to all cultures, and secondly, a process of evaluating their interpretation in the current situation. As Toiskallio (2006, 2007; see also Robinson, 2007) argues, there will not be a uniform understanding of ethics, and it is most probable that Heinonen's ideation will remain in the state of utopia. But the undisputable necessity of ethical discussion in the military everywhere (e.g., Reichberg et al,

2006; Toiskallio, 2006; Värri, 2007; Verweij, 2007) obliges the educators to seek an understanding about how to put this ongoing process into action in teaching.

The teaching of ethics is not only translating a collection of codes and roles to rules of engagement or transmitting them to soldiers, it should be empowerment and activation of people to become aware of ethical problems (Toiskallio , 2007; van Baarda & Verweij, 2006) requiring an integrative curriculum (Raviv, 2005); ethics should be integrated as a part of regular military life (Robinson, 2007; van Baarda & Verweij, 2006). Having normative rules connected to rewards and punishments will not be enough for a person to become a correctly thinking and acting human (Martinelli-Fernandez, 2006), as this does not develop a skill of solving problems when there are conflicts between the rules (virtues). To educate in the area of ethics is to touch something which derives from the very sphere of the individual's lifelong experiences, knowledge, attitudes and perceptions (Toiskallio, 2007). Robinson (2007) even suggests that actual formal training may not achieve much in the field of ethics.

Prior (2000) writes that to be a modern warrior one must be educated in such a way that he/she will feel the conflict between the morality of decency and that of war, i.e., to be able to feel guilt about his/her actions in combat.[2] Martinelli-Fernandez (2006) suggests that the goal of moral education (we see the concept *moral agency,* as put forward by Martinelli-Fernandes, closely related to good ethical conduct) could be defined through the idea of Kantian autonomy as a sensitivity and an awareness about how to achieve good ethical conduct. Furthermore, she writes that training in ethics towards the Kantian autonomy will help the trainees to achieve a level of practical reasoning that will enable them to meet challenging situations by governing themselves with certain laws and principles. The distinguishing feature of a mature moral agent would be confirming to moral principles *voluntarily and for their own sake.* For example, ethical military leaders are aware of and capable and willing to use their reasoning to override unethical orders (Martinelli-Fernandez, 2006; van Baarda & van der Heijden, 2006). In Kohlbergs model (according to Robinson, 2007), ethical education should aim towards the post-conventional level, where individuals use their own reasoning to define right from wrong on universally good ethical principles, because they themselves have *chosen to do so.* This is very similar to what van Baarda & Verweij (2006) write about the desired outcome of education: "... people who are clearly loyal and disciplined, but who retain a sovereign mind."

On the basis of the theoretical premises presented above, for the sake of sound education we now ask that since ethical behavior in, e.g., decision-making

---

[2] Prior defines the morality of decency thus that its fundamental concept is universal respect for all human beings as moral agents. The morality of war, according to Prior, is that, e.g., survival, mission or duty are of overriding importance.

is such an intimate and even unconscious process, what requirements will this set to the methodology and techniques of teaching?

Conceptualizing learner: constructivism and self-regulation in learning

Self-regulated learning concerns the application of a certain model of regulation and self-regulation to issues of learning, and therefore instructional implications are relevant. Pintrich (2000) presents that the multitude and overlapping models of self-regulated learning (e.g., social cognitive conception presented by Zimmerman, 1989) share common ground in their profound premises. These premises can be condensed (Pintrich, 2000) in the form of general assumptions as follows:

1. The learner is considered to be an active and constructive participant in the learning process.
2. The learner can monitor, regulate, and control certain aspects of his/her own abilities and those of the environment.
3. There is some type of criterion against which the process is compared.
4. Self-regulatory activities are mediators between personal and contextual characteristics and actual achievement.

The concept of self-regulating learning is a part of the general concept of constructivist learning theory (Brooks & Brooks, 1995; Savery & Duffy, 1996; Tishman, Perkins & Jay, 1995). The theory of constructivism is one of the most influential educational theories in the 20th century. Its elements and principles are of utmost relevance to teaching military ethics.

The leading principles that should be adapted to working with officers on questions of attitudes, moral, and ethics are:

1. Effective learning is based on active conceptualization of information: learning is an active process in which one uses sensory input and constructs meaning from it. Existing knowledge of the world is used to understand new information.
2. Learning is a social process, and this is mediated by personal and cultural points of view, as our learning is intimately associated with our connection with other human beings, our teachers, our peers, our family, as well as casual acquaintances.
3. Learning is accelerated through dialogue, feedback, and discourse.
4. There is a need for authentic tasks that reflect the complexity of the real and relevant world.
5. Encouraging alternative views, mental inquiry, and testing ideas are a part of elaborating critical thinking and building a challenging character.

6. Learning is contextual: we learn in relationship to what we know, what we believe, our prejudices, and our fears.

The learning process can be reified into four phases, in which the self-regulating processes themselves may take place in four areas. These areas are cognition, goal orientation, behavior, and context. As this study does not concern particularly the area of motivational constructs in the learning process, we do not discuss the area of goal orientations; for motivating students and establishing interest is seen as a natural part of teaching. There is evidence that intrinsic value is related to self-regulation, i.e., the student's involvement in self-regulated learning is closely tied to his/her beliefs on how interesting and worth learning the tasks are (e.g., Pintrich & De Groot, 1990). We have also chosen to omit the area of context from this article as the discussion will take place at a general level.[3] The areas of regulation of cognition and behavior are presented in detail in Table 1, as they focus more on the aim of this study to address the methodology of teaching. The cells of Table 1 represent how the phases may be applied in the different areas. Although the nature of any learning process cannot really be compressed into an exact mould, this simplification gives an idea of how to form further working definitions to discuss this matter. The phases are not clearly or strictly separated and their succession may not be linear, but rather dynamic and interlacing. Also, the borders of the areas of regulation are somewhat fuzzy (Pintrich 2000).

Table 1. Phases and areas of self-regulating learning

Phases	Areas of regulation	
	Cognition	Behavior
1. Forethought, planning and activation	Target goal setting Prior content knowledge activation Metacognitive knowledge activation	Time and effort planning Planning of self-observations of behavior
2. Monitoring	Metacognitive awareness and monitoring of cognition	Awareness and monitoring of effort, time use, need for help Self-observation of behavior
3. Control	Selection and adaptation of cognitive strategies	Increase/decrease effort Persist, give up Help-seeking behavior
4. Reaction and reflection	Cognitive judgment Attributions	Choice behavior

(modified from Pintrich, 2000)

---

[3] Pintrich (2000) notes that not all models of self-regulation include this area, as it can be seen as an external one, whereas for Zimmerman (1989) activity becomes *self-regulated* when environment-initiated strategies come under personal processes.

Self-regulated learning is defined as an active, constructive process where the learners set goals for their learning and try to monitor, control and regulate their learning guided and constrained by their personal learning goals (Pintrich, 2000; Pintrich & De Groot, 1990; Zimmerman, 1989). With the elaborations on the conception of the learner, the learning process, and the nature of ethics, we will discuss how the case study method is consistent with the idea of promoting the learning of ethics.

Cases to learn from – a general introduction to the case study method

Case studies are an ideal method when a holistic, in-depth investigation is needed. Case study research excels at bringing us to an understanding of a complex issue or object, and it can extend experience or add strength to what is already known through previous research. Case studies emphasize detailed contextual analysis of a limited number of events or conditions whose relationships and context are not clearly evident, and in which multiple sources of evidence are used (Yin, 1984).

An effective teaching case is actually a story that describes or is based upon real world events or circumstances. This story should carry specific learning objectives, and demands profound and deep analysis (Lynn, 1999). Cases are narrative accounts of actual, or realistic, situations in which policy makers are confronted with the need to make a decision. Cases supply students with information, but not analysis. Case method teaching is a group *enterprise* in which the emphasis is on self-discovery by the class, working together with the guidance of the instructor and producing perceptions, solutions, and question marks where needed. Barnes, Christensen and Hansen (1994) refer to the case study as an "account of events that seem to include enough intriguing decision points and provocative undercurrents to make a discussion group want to think and argue about them". Complex and information-rich cases depict incidents that are open to interpretation – raising questions rather than answering them, encouraging problem solving, calling forth collective intelligence and varied perspectives (Hutchings, 1993; cited by Christudason, 2003).

Military ethics is one of the most complex issues to learn. Developing a commander's ethical awareness means making him capable of distinguishing actual from declared values, of engaging in self-criticism, and of changing. He should internalize and pass on to others universal and professional values, and still give a good enough response to four crucial, inherited tensions: performance of the mission according to its aim, preserving his troops' lives, preserving the lives and dignity of innocent civilians, and preserving the lives of citizens (Raviv, 2007).

Fleck (2005) writes that to be able to recognize whether something is wrong with respect to ethics is the first condition for ethical behavior. Ethical dilemmas are profoundly complicated deductions to be solved in a jungle of controversial goals and ambiguous directions to follow, and as Raviv (2005) writes, there are no trivial answers to true ethical problems (see also van Baarda & van der Hejden, 2006). Readiness to initiate discussions and disseminations with colleagues of ethical codes and their conduct (Talerud, 2007) as well as constant debate between cognition and affection in both intrapersonal an individual levels (Raviv, 2005) lay at the core of ethical education. The ability of self-reflection as a willingness and consciousness for self-forming activity in order to re-evaluate one's relationship with other humans (Talerud, 2007; Toiskallio 2006, 2007) is the key objective of ethical education.

To promote this understanding and the ability to see the dilemmas and speculate on the alternatives, we need to build a curriculum that can to cross the barriers. The case study method is an excellent one to face those challenges. We have chosen case studies extracted from real-life situations as examples for several reasons; although a holistic view of education is needed to understand the development of ethical competence, some form of formal training is needed to create common a ground and to avoid the spirit of elitism (Robinson, 2007) as well as to ensure that the ethical dilemmas faced are consistent with the environment the trainees are likely to face.

The contents of cases are not very valuable to students without deliberate guidance on the subject. If the examples are put to pedagogical use without deep knowledge of the context it must also be admitted that written cases are usually simplified 2nd or 3rd grade interpretations of very complex situations, where all the facts of the case may not even be known. For this reason it can be said that solutions offered by scholars to such dilemmas may sometimes carry nothing more than the burden of their makers' prejudice. It is also plausible that the trainees cannot make a distinction between true discussions on a sensitive and complex matter and a politically and subjectively twisted scheme. Therefore it is *the evaluation and critical dissemination* of the cases that will promote ethical competence. Martinelli-Fernandez (2006) writes that Kant's steps of moral education remain valid and states that "through an agent's assessment of the actions of another one is set on the path of moral agency". Didactically, the instructor should foremost set the conditions for the students to develop a habit of moral assessment. The expertise of the instructors is also a prerequisite for success, as case-studies are only effective if the instructors themselves have knowledge of what the case studies are meant to demonstrate (Robinson, 2007).

The Dutch approach: the flow model to reach the decision

In the Dutch example (see van Baarda, 2006a; van Baarda, 2006b) of forming a moral judgment, a so called dynamic model (flow model) is introduced. In the flow model, solution is sought via a recognition and re-definition of a problem through a path to knowledge, and consideration of the goals and means is related to the problem. This process is illustrated in Figure 1 and broken into five steps as follows (the superscripts refer to Figure 1):

1. From perceiving a problem^A to describing the facts^B
2. From observation^B to interpretation^C
3. From preference^C to feasibility^D
4. From the path of choice^E to the decision^A : matching the path of knowledge and the path of choice.(van Baarda, 2006a)

Figure 1. Illustration of the Dynamical Model.

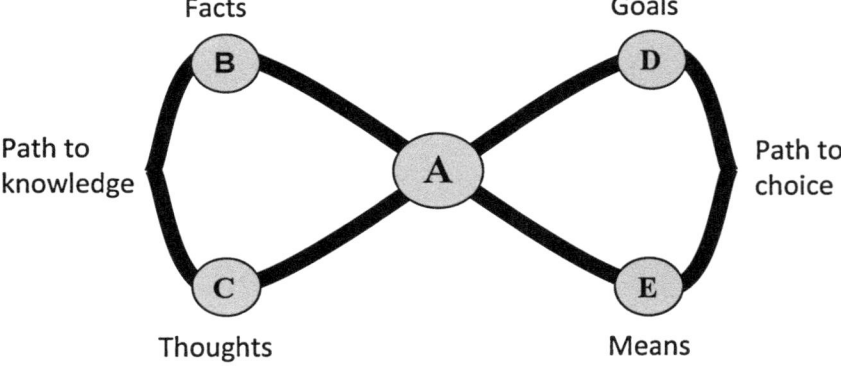

(van Baarda, 2006a, p. 285)

It should be noted that although this may seem a very simplified and, to some scholars, quite mechanistic presentation, van Baarda (2006a) notes that it has been reified for the purpose of readability, and the presentation is given without scientific discussion of the background of the model. Van Baarda (2006a; 2006b) also clearly states that the focus is that the flow model should not be seen as a formula of some kind or a procedural directive: "A well balanced judgment is seldom instantaneous: it is preceded by a whole process. A judgment process rarely follows a straight line; on the contrary, it is a dynamic process". The unique nature of this model is revealed by comparing it to other decision making models, which van Baarda refers to as rational theories, so that the flow model takes the person making the decision into the account, i.e., in training situation students are involved in the situation instead of only commenting and evaluating it from the outside (van Baarda, 2006a; 2006b).

The instruction using the presented model is constructed of certain phases and exercises and emphasizes the social and discussion skills needed to deal with emotionally charged subjects. The purpose of this method is to teach how a morally responsible decision can be made, and it is of complementary nature to normal decision-making procedures, as concern is expressed on whether resistance will be met if the model is (mistakenly) identified to be a consensus model not fitting in the chain of command (van Baarda, 2006b.). Even though it is meant that the process and the skills calling for the reflection of one's own actions and premises would be the focus of this method, it is quite obvious that if the manual is followed to detail, i.e., by disseminating any given case according to the five steps described above, the instructor will be required to have notable skills himself so that the training will not succumb to rote learning of yet another procedure. The Dutch manual also presents a collection of hypothetical dilemmas to be used in training; the cases are written with a lot of information and they only lay out the situation; there are no solutions or added guidelines to discussions (Springer, 2006).

It needs to be noted that this discussion is not to be taken as a valid critique based on a thorough analysis of the fine Dutch volume, as it carries highly structured theoretical discussions in profound articles creating a framework much deeper and wider than is practical to take into account in this study. Every attempt to contribute to teaching methodology is most welcome, but also likely to highlight the numerous difficulties and ambiguities of education in the field of ethics.

The Canadian way: the value based approach to decision making

The Canadian Forces base their ethical considerations on a construct called value-based decision making. In this framework it is seen that the military ethos is comprised of a) Canadian military values that are elaborated from contributions of b) beliefs and expectations about military service that set the conditions and values essential for military effectiveness, and c) Canadian values, expectations and beliefs that provide a philosophy of service. Professionalism is shaped by the ethos and therefore the ethos governs the conduct (Canadian Defence Academy, 2006).

The Canadian instructors' manual introduces a sequence of discussion guidelines (also referred to as *steps*) to be employed in training (and supposedly, in a way, in real events) that include a) assessment of the situation, b) ethical considerations, and c) options and risks. *Assessment of the situation* (here) is a general summary in which facts and perceptions of the situation, as well as other issues, including implicit and personal and environmental factors, are taken into account. *Ethical considerations* include identifying the relevant ethical principles of the Canadian Forces and the listed ethical values that shape the conduct of its members. Via these considerations, the type of ethical dilemma is determined. In the step of *options and risks* a variety of possible courses of action are evaluated and reflected. These guidelines or steps are followed by committing to action, wherein the student must choose a solution he/she has outlined in the process, or by combining aspects from several options. The case studies presented in the instructors' manual all include tailored contents on these guidelines (ibid.).

Practical application of this method is introduced with advice that the provided responses are only "possible", in other words, the instructor can use them as a starting point for discussions. The authors warn that effective solutions are not limited to this selection. Also, the users of the instructors' manual are invited to modify the given answers, guidelines and aspects of the cases. Both the Dutch and the Canadian argue on the behalf of personal involvement of students. A fundamentally important feature is that the student must position him-/herself as the one responsible in the event (ibid.).

The case studies in the Canadian manual are presented in a very condensed and template-like manner. An interesting, though not essential matter concerning pedagogical implications, is that whereas the Dutch presentation clearly states that their dilemmas are hypothetical and only carry coincidental similarities with real events, the Canadian explicitly point out that their cases are transcribed from real-life situations (Canadian Defence Academy, 2006; Springer, 2006).

The chances are high that these exemplary guidelines and their contents might be taken as an answer at face value or even as rules of thumb for what is really right. This, as discussed earlier in this chapter is a challenge when using such materials as pedagogical materials. The use of case studies, i.e., dilemmas, requires the teachers not to oversimplify the level of unambiguous problem-solving, but also to make sure that profound abilities are developed.

The Israeli way: the multiple approach attitude

The IDF colleges have developed a holistic approach for teaching military ethics. The curriculum was developed as a response to the unique and complex security environment Israel lives in, confronting conventional and non-conventional threats, and of course, the threat of in-house terror. Israel is the only democracy in the world that faces an enormous terror threat: in term of endurance, span, and intensity. Israeli commanders should be prepared to apply military ethics when confronting regular Arab militaries, homicide-suicide terrorists, terror-oriented guerrilla forces, and even when operating in front of Israeli citizens when illegal settlers have to be evacuated.

The IDF colleges have developed a learning approach containing multi-dimensional pedagogical and contextual elements that are based on ten arguments:

a) The professional angle

The IDF puts a specific emphasis on developing the soldier's proper behavior and his professional identity. This is done under the assumption that each practice in valuable subjects is a step up in the professional area. The IDF code and concept of military ethics is of a practical ideal of our behavior (Kasher, 1997). Professionalism means knowing things you did not know earlier in their depth, to develop unique skills, and to dedicate attention to implicit dimensions of the profession. Each profession needs to develop its own professional identity, not only universal moral values. Therefore great emphasis is laid upon the idea of a military within the democratic state, and constraints are put on the military commanders and soldiers as a result.

b) Cases that touch the three interfaces

When developing ethical awareness it is important to notice the three main entities with which the military professional interacts: the people – the basic body for which he works, i.e., the democratic state and all its citizens; colleagues – commanders, subordinates, the unit; and obviously – the enemy. The IDF Spirit deals with all three, and sets principles that give the right value for each interaction. However, during the years there has been a different emphasis as a consequence of actual events. For example, let us take a unique event: during 2005, disengagement from the Gaza Strip took place, meaning the evacuation of 25 settlements in the Gaza Strip and the Northern West Bank. This created a dramatic test to the Israeli society and democracy and put a complex challenge for Israel's interagency security cooperation. Therefore there was a need to design a case study-based curriculum that sets up the experienced intellectual and emotional dilemmas concerning the questions of: How will the IDF succeed in executing a law enforcement mission? How will the military face violence from Israelis and still keep the IDF Spirit? How will we deal with implications of refusal? Hence, the colleges should constantly develop cases that reflect the complex, actual reality and invite the officers to present their cases concerning these subjects.

c) Different analysis of fighting terror and high intensity conflict

The Spirit of the IDF is the identity card of the IDF values, which should stand as the foundation of all of the activities of every IDF soldier, on regular or reserve duty. It contains universal values such as *tenacity of purpose in performing missions and drive to victory*, as well as *responsibility, personal example, purity of arms* and so on. However, it seems that the conditions of the fight against terror are essentially different from the conditions that are assumed to exist in the classical war or the law enforcement paradigm (Kasher & Yadlin, 2005). Therefore a third model, as Kasher and Yadlin claim, was needed.

In the Post Cold War, conventional conflicts between sovereign national states have been replaced by civil, religious, gang, or ethnic wars. They are marked by diffuse power structures, missing force monopolies and often changing boundaries between enemies and allies, by asymmetrical warfare using civilian population as basic resource, by migrations, and by humanitarian catastrophes (Kaldor, 1999).

The soldier's classic actions are aimed at attacking and destroying an enemy, if necessary by all means. Ambiguous situations are disconcerting for the soldier and often provoke falling back on trained, reflexive behavior (Haltiner, 2003). Therefore there is a need to apply different principles of ethical consideration. Those principles should reflect real and hypothetical cases, and require application of principles such as self-defense duty, military necessity, or principle of distinction, for example.

When dealing with terrorist entities, which by their definition conceal themselves amongst civilian population, and whose target of harm is the civilian population, one needs an elaborated angle to deal with the challenge. Therefore, the curriculum should include historical and hypothetical cases that focus on these different issues in order to prepare the officers for the future.

d) Four contradictions

When fighting terror, many ethical questions rise in calling the "right" decision. For example, when is it justified to impose a siege around a town when we know for sure that a suicide bomber is about to leave it, carrying explosives on his body in order to commit homicide in a crowded bus? When is it justified to thoroughly check all passengers at a road block? Every action the commander chooses has its own price and further consequences. Each choice has to take into consideration questions of the commander's responsibility for the soldiers' lives, the obvious need to complete the mission, the keeping of human dignity and the daily routine of civilians on the other side, and sustaining your own civilians' security by thwarting all enemy efforts to disrupt the normal way of life in Israel. Almost every choice encloses harm in one or more dimensions. This insecure reality, together with changes in the society's values concerning canonical organizations, requires brave treatment of war moral, ethics, and leadership questions.

e) The ethical chapter and the embedded holistic curriculum

The basis of military ethics curriculum is grasping the subject consisting of supplementary contradictions and a combination of abstract ideas which are connected to the military's duty in a democratic country. Also the portrayals of practical principles that make cognitive tools and are part of the professional practice are needed (Raviv, 2005). Another contradiction is the need to give simple and practical answers in spite of the complex and dynamic world in which the officers meet new, unfamiliar situations. In this ever-changing reality, there is an ever-growing difficulty to provide the officer with a system of orders and procedures that will resolve any problem that might occur. To this reality, the professionals must be supplied with cognitive and analytic tools that are wide-ranging and smart, and not technical rules of "dos and don'ts". All this requires an integrative curriculum that includes ethical dealing as a separate or intertwined chapter, as a formal lesson, as a feedback and a systematic appraisal of students, as the subject of the staff's and of external lecturers' teaching, as an area through which the IDF's activity today and in past wars is to be inspected, and as a criterion by which the college appraises the staff, the course, or the individual. The ethical prism will merge when discussing combat, but also when talking about the force-building routine.

f) The learner's experience

All colleges' officers are mature and adult learners. Majors, lieutenant colonels and colonels, they all possess a rich source of expertise, experience and self guidance capabilities, as well as skepticism. They are instrumentally interested in the teaching topics, i.e., they are purpose oriented. They are critical and creative thinkers who can adapt and thrive in ambiguous and ever-changing environments. Therefore, the topics should be relevant to their everyday problems, and connected to the role of self-concept.

Thus, the case study method, as a part of varying teaching methods, is highly suitable. It promotes the ability to listen, respect other considerations, and stand behind your own views. It makes it possible to change an attitude, alter a view, improve the ability to contribute to the decision making process as a social or political process, it can strengthen the will and ability to promote change, and solve professional problems. The method assists in providing a pedagogical atmosphere of openness, authentic argumentation, and criticism.

g) Instruction provided by the staff and field commanders

The educational process of military ethics is led by the commanders and the academic professors. The professors set the theoretical base and the fundamental concepts, but the commanders are responsible for relating it to professional combat reality. The commander is the subject of identification; he is an example for soldiers and creates stimuli and learning experiences. His role is to provide ethical interpretation at every opportunity, not only in the ordered chapter of the program, but in the analysis of military topics, combat drills, military history, etc. It is imperative that the staff itself has a thorough conceptual framework for ethic evaluation, and the skill to encourage reflective thought (Raviv, 2005). The staff must professionalize and familiarize itself with the considerations behind the Spirit of the IDF, the essence of the military in the democratic state, and international law. The staff should take advantage of different events in the course and design them into a critical ethical event that creates an emotional stimulus for further strengthening of the learning process.

However, working with cases sets a psychological challenge for the instructors. It demands a thorough understanding of the case and its alternatives, skills of discussion facilitation, an ability to clarify and summarize without controlling the dynamic process, and it demands an ability to learn from one's class, too.

h) The ideal phases of learning and teaching with cases

Working with cases has four stages: designing the case, preparing the group, discussions, and summary. Sometimes it is the students themselves who present their cases, sometimes it is the instructor who chooses the specific case. Every effective facilitation of a case should contain the following analytic stages (Lynn, 1999):

a) Understanding the facts: Who? What? When? Where?
b) Analyzing the facts: Why? How come?
c) Mapping the core challenge: So what? What is the meaning of all this?
d) Action: What would *you* do?
e) Raising assumptions: What would have happened if?
f) Forecasting: What will happen as a consequence?
g) Lesson learned and conclusions: This is an example of [...]? What is the meaning of this example? Where should it lead us?

An ideal fruitful case discussion should rely on these phases, as well as on integration and a summary of all the topics that were raised in class. Special attention may be given to relevant points that were not mentioned, and the instructor should trigger a short discussion on the reasons why these topics were ignored.

i) Types of cases

There are five types of cases (Lynn, 1999):
a) Cases that require a decision in the event that there are constraints or vague data
b) Cases that demand a policy making and a framework of conceptualization
c) Cases that demand a problem definition
d) Cases that call for implementation of an idea or theory
e) Demonstration of a historical case – which resembles a frontal lecture.

The IDF colleges try to represent a blend of all five, and maybe tend to use more the first type. Because of the complex security reality in Israel, there will always be a need for a transformation of learning from class to the battlefield. All commanders constantly make ethical decisions; therefore, they should be instructed in class.

j) Using tactical and operational decision making cases

It is understood that national defense students usually tend to analyze cases that deal with national flair and dimensions, in the context of political-security consequences, inner-society issues, and civil military relations. The younger officers, however, tend to analyze more tactical cases, which are closer to their world and reality as unit commanders.

However, although the case study method is common in the IDF colleges, a constant pedagogical elaboration is needed, as well as constant developing of new cases. Every instructor has his own teaching style, and a lot of effort should be invested into the general code and the direction the college as a whole should be steering.

Summary

Military ethics is one of the most complicated subjects professionals need to learn. Because of the aspects mentioned above, touching the learners' hidden agendas can influence the will to change and elaborate normative thinking. Teaching with cases can upgrade this influence and invite the learner to explore his own self and the military institution in a journey of elaboration and self inspection. Combining cases with a solid theoretical background, as well as building a multi-dimensional approach can minimize the danger of over-generalization and over-simplification from a single specific event.

The examples of different national points of view, related to the role and realities of armed forces of each nation, seem to suggest that although the case study method, adopted in different forms, is a most promising choice to promote ethical competence, there may be interesting if yet implicit features to be discovered. As it is, the case study method requires very much from the narratives or stories presented, i.e., these requirements fall on educators engaged in creating and evaluating the materials. All the approaches discussed clearly indicate that the role of the teacher is far from the traditional image of a schoolmaster, as it is from the traditional image of a military instructor drilling his/her subordinates. The teaching of ethics, and especially the case study method, demands sophisticated perceptions of learning and the learner, skill in the guidance of learning, social skills combined with a right attitude, and deep knowledge of the phenomena under dissemination.

Therefore, although the case study method is an excellent choice in the field of teaching ethics, more detailed information about the interaction between several aspects are needed for further development of applications. For example, how are the instructor's personal methodological elaborations discussed among the educators and the students? Do the personal epistemologies of the educators and the students predict their perceptions of the actual learning sessions and the outcome of training? How are the educators educated prior to their employment in the field of teaching ethics? What kind of pedagogical simulations fully prepare them to facilitate a case discussion?

There is also a need for further research on the possible models marking the proportion between cases and the general curriculum. What is the vital knowledge needed prior to learning with cases? What correlation should there be between the student's learning styles and the case method?

The meanings and relations of motivational constructs and the contexts of learning need to be clarified to gain a more coherent idea of how people learn ethics. Further research on this area should utilize the possibilities to gather empirical data during ethical training in order to address the challenges of military pedagogy when the need for ethical education is becoming more and more crucial.

Bibliography

Barnes, L.B., Christensen, C.R. & Hansen, A.J. (1994). *Teaching and the Case Method: Text, Cases, and Readings (3rd ed.)*. Boston: Harvard Business School Press.

Brooks, M. & Brooks, J.G. (1995). *The Case for Constructivist Classrooms*. Alexandria: ASCD.

Canadian Defence Academy (2006). *Ethics in the Canadian Forces: Making Tough Choices*. Instructor manual.

Christudason, A. (2003). *The Student's Role in Case-based Learning*. Successful Learning, Vol. 10. Retrieved from http://www.cdtl.nus.edu.sg/success/sl10.htm.

Fleck, G. (2005). Teaching Ethics: A Psychological View. In Micewski, E. & Annen, H. (Eds.) *Military Ethics in Professional Military Education – Revisited* (pp. 65-74). Frankfurt: Peter Lang.

Fry, D. (2005). *The Human Potential for Peace. An Anthropological Challenge to Assumptions about War and Violence*. New York: Oxford University Press.

Haltiner, K. (2003). Athens versus Sparta – The New Missions and the Future of Military Education in Europe. In H. Kirkels, W. Klinkert & R. Moelker (Eds.), *Officer Education: The Road to Athens!* (pp. 177-192). Breda: Koninklijke Militaire Academie.

Heinonen, R. (2002). Values Memory, Global Ethic and Civil Crisis Management. In Toiskallio, J., Royl, W., Heinonen, R. & Halonen, P. (Eds.). *Cultures, Values and Future Soldiers* (pp. 67-96). Helsinki: Edita.

Hutchings, P. (1993). *Using Cases to Improve College Teaching: A Guide to More Reflective Practice*. Washington, DC: American Association for Higher Education.

Kaldor, M. (1999). *New and Old Wars. Organized Violence in a Global Era*. Stanford: Stanford University Press.

Kasher, A. (1997). *Military Ethics*. Tel Aviv: Ministry of Defence-Publication House.

Kasher, A. (2003). Professional Ethics. In G. Schefler, Y. Achmon & G. Weil (Eds.), *Ethics in psychology* (pp. 15-29). Jerusalem: Hebrew University Magnes Press.

Kasher, A. & Yadlin, A. (2005). Military Ethics of Fighting Terror: An Israeli Perspective. *Journal of Military Ethics, 4*, 3-32.

Lynn, L. Jr. (1999). *Teaching and Learning with Cases*. New York: Chatham House Publishers.

Martinelli-Fernandez, S. (2006). Educating Honorable Warriors. *Journal of Military Ethics, 5* (1), 55-66.

Micewski, E. (2005). On the Philosophical Framework for the Ethical Debate in Professional Military Education. In Micewski, E. & Annen, H. (Eds.), *Military Ethics in Professional Military Education – Revisited* (pp. 13-21). Frankfurt: Peter Lang.

Micewski, E. & Annen, H. (2005). Preface. In Micewski, E. & Annen, H. (Eds.), *Military Ethics in Professional Military Education – Revisited* (pp. 7-9). Frankfurt: Peter Lang.

Pintrich, P. (2000). The Role of Goal Orientation in Self-Regulated Learning. In Boekaerts, M., Pintrich, p. & Zeidner, M. (Eds.), *The Handbook of Self-Regulation* (pp. 451-502). San Diego: Academic Press.

Pintrich, p. & De Groot, E. (1990). Motivational and Self-Regulated Learning Components of Classroom Academic Performance. *Journal of Educational Psychology, 82* (1), 33-40.

Prior, W. (2000). Saving Private Ryan and the morality of war. *Parameters, 30* (3), 138-146.

Raviv, A. (2005). Teaching Values for Military Professionals. In Micewski, E. & Annen, H. (Eds.), *Military Ethics in Professional Military Education – Revisited* (pp. 55-59). Frankfurt: Peter Lang.

Raviv, A. (2007). Command as the Profession of Military Officers: Concepts, Components, Education and Training Principles. In Toiskallio, J. (Ed.), *Ethical Education in the Military: What, How and Why in the 21th Century?* (pp. 87-99). Helsinki: ACIE Publications.

Reichberg, G., Syse, H. & Begby, E. (Eds.) (2006). *The Ethics of War. Classic and Contemporary Readings*. Oxford: Blackwell Publishing Ltd.

Robinson, P. (2007). Ethics Training and Development in the Military. *Parameters, 37* (1), 23-36.

Savery, J. & Duffy, T. (1996). Problem based Learning: An Instructional Model and its Constructivist Framework. In Brent G. (Ed.), *Constructivist Learning Environments: Case Studies in Instructional Design* (pp. 135-148). Englewood Cliffs: Educational Technology Publication.

Springer, S. (2006). Dilemmas for Use during Instruction. In van Baarda, T & Verweij, D. (Eds.), *Military Ethics. The Dutch Approach* (pp. 345-354). Leiden: Koninklijke Brill NV.

Talerud, B. (2007). Ethos in and Ethics for Today's Armed Forces in the Western Societies. Why bother about military ethics and ethos? – Setting the stage. In Toiskallio, J. (Ed.), *Ethical Education in the Military. What, How and Why in the 21st Century?* (pp. 63-48). Helsinki: National Defence University.

Tishman, S., Perkins, D & Jay, E. (1995). *The Thinking Classroom: Learning and Teaching in a Culture of Thinking*. Needham Heights: Allyn & Bacon.

Toiskallio, J. (2006). Etiikka sotilaspedagogiikan ytimessä. [Ethics in the core of military pedagogy] In Huhtinen, A.-M. & Toiskallio, J.(Eds.), *Maanpuolustuskorkeakoulu – kehittyvä sotatieteellinen yliopisto. [National Defence College – the Developing Military-scientific University]* (pp. 121-150). Helsinki: Edita.

Toiskallio, J. (2007). Introduction: Edifying Military Ethics. In Toiskallio, J. (Ed.), *Ethical Education in the Military: What, How and Why in the 21th Century?* (pp. 11-27). Helsinki: ACIE.

Tomasello, M. (1999). *The Cultural Origins of Human Cognition.* London: Harvard University Press.

van Baarda, T. (2006a). Forming a Moral Judgement Using a Dynamic Model. In van Baarda, T & Verweij, D. (Eds.), *Military Ethics. The Dutch Approach* (pp. 279-298). Leiden: Koninklijke Brill NV.

van Baarda, T. (2006b). Manual for Instructors: Forming a Moral Judgement Using a Dynamic Model. In van Baarda, T & Verweij, D. (Eds.), *Military Ethics. The Dutch Approach* (pp. 299-330). Leiden: Koninklijke Brill NV.

van Baarda, T. & van der Heijden, P. (2006). Ethics and Dilemmas in the Royal Netherlands Air Force. In van Baarda, T & Verweij, D. (Eds.), *Military Ethics. The Dutch Approach* (pp. 149-172). Leiden: Koninklijke Brill NV.

van Baarda, T. & Verweij, D. (2006). Military Ethics: Its Nature and Pedagogy. In van Baarda, T & Verweij, D. (Eds.), *Military Ethics. The Dutch Approach* (pp. 1-24). Leiden: Koninklijke Brill NV.

Värri, V-M. (2007). Some Problems of Ethics in Military Education. The Question of Ethics in the Military Space. In Toiskallio, J. (Ed.), *Ethical Education in the Military: What, How and Why in the 21th Century?* (pp. 31-42). Helsinki: ACIE Publications.

Verweij, D. (2007). Military Ethics: A Contradiction in Terms? In Toiskallio, J. (Ed.), *Ethical Education in the Military: What, How and Why in the 21th Century?* (pp. 43-62). Helsinki: ACIE Publications.

Yin, R. K. (1984). *Case study research: Design and methods.* Newbury Park: Sage.

Zimmerman, B. (1989). A Social Cognitive View of Self-Regulated Academic Learning. *Journal of Educational Psychology, 81*(3), 329-339.

Hubert Annen

Coaching for Military Personnel – From the Idea to the Implementation

Introduction

Working as a military professional has always been demanding. In Switzerland for instance, the legal employment conditions of officers and NCOs demand working hours depending on current requirements, i.e., without upper limits. Moreover, the budget for military defense in Switzerland has been substantially reduced and several reforms of the armed forces and the administration have yielded changes in duties and responsibilities. Thus, professional officers and NCOs are required "to do more with less", while at the same time job security and career opportunities are reduced.

Since in a governmental organization the financial scope for bonuses or incentives is limited, one has to consider other ways to keep the military professionals' job satisfaction and commitment on a favorable level. It is known that factors such as social support, appreciation and understanding foster self-esteem and accordingly well-being at work. The project "*Coaching for military personnel*" explicitly takes these aspects into account. Professional officers and NCOs with specific requests and problems in terms of job, career, work-life balance, etc., can contact experienced colleagues that are specifically trained as coaches.

Starting point

A characteristic feature of the Swiss Armed Forces is that the majority of officers and NCOs are members of the militia. Accordingly, professional officers and NCOs as well as professional military personnel account for only about 3% of the total Swiss Armed Forces. Hence, professional military personnel form the mainstay of the armed forces in terms of know-how and skills, serving as a gauge for any military instruction, command and education.

The high standards applied to professional military personnel are reflected in the systematic selection process. Prospective career officers, for instance, are required to pass an assessment centre where social skills are systematically monitored and appraised and mental abilities are tested (Annen, 2007). This assessment yields a valid prediction with regard to the ambitious course of studies leading to the attainment of an internationally recognized B.A. degree from the ETH Zurich (Annen & Eggimann, 2006).

In practice, career officers serve as company commanders in basic military training or cadre training and as instructors or coaches of militia cadre. Later on, they assume challenging leadership positions, take responsibility for projects or conduct the educational process in specific fields. The prevailing principle is that working hours are flexible and dependent on the current necessities of the troops. According to a recent survey the average weekly working time is close to 56 hours (Stocker, Jacobshagen, Semmer & Annen, 2010). As is generally known, a challenging occupation does not necessarily lead to discontentment, provided that framework conditions are appropriate (e.g., Herzberg, 1974). From the military professional's viewpoint, however, there have been significant impairments within the past ten years. In particular, the two reforms of the armed forces in 1995 and 2004 must be mentioned here. Besides creating a sense of uncertainty among the personnel – a fact that is often observed during such endeavors – these reforms in many cases entailed a higher workload and more stringent requirements, yet fewer career opportunities. Moreover, insurance and fringe benefits were curtailed and expense regulations were tightened.

While job satisfaction of the military professionals remained high nonetheless, the loss of certain framework factors yielded a significantly reduced emotional commitment to the organization (Annen, 2004). As a consequence, the number of resignations soared while job applications slumped. In light of this evidence, concrete measures were taken in order to counteract these tendencies.

Various task forces were formed to address specific questions with the aim of improving the job image, general conditions, and the motivation of the military professionals. Among other things a temporary wage increase was granted as a sign of awareness of the problems and of the determination to tackle them. In addition, visible, concerted efforts such as special leadership trainings were undertaken in order to improve social skills and management culture. The Armed Forces Command was relentless in emphasizing the fact that the human shall take centre stage in all respects. Such measures prove successful if the person concerned perceives them as a sign of appreciation (Semmer, Tschan, Meier, Facchin, & Jacobshagen, 2010). However, the sustainability of the prescribed training remains questionable (e.g., Bergmann, 2000). Moreover, affirmative statements and promises on the part of the Armed Forces Command bear the risk of being counterproductive. That is, if a professional military feels neglected in his individual circumstances and lacks support, he might gain the impression that all of these efforts are no more than lip service intended to appease the critics.

The idea

Specific investigations have shown job discontentment to crucially depend on the perceived leadership ability of the line supervisor (Gutknecht & Krautz, 2004). Conversely, the appreciation received is a significant factor of satisfaction and stress mitigation. This applies in particular when the workload is heavy (Stocker et al., 2010).

In practice, reforming the armed forces and therefore altering structures within has resulted in a high personnel turnover. Consequently, many young career officers and NCOs lack an experienced and reliable contact person. Then again, professional military personnel over the age of 50 years with few real career prospects often admit to mental resignation while some even go so far as to claim they count the days until retirement. Most worrisome to them is the fact that they often feel pushed aside and given insignificant positions. They lack the opportunity to share their valuable propositions and experience with others.

An obvious benefit to both parties would be if the experienced professional cadre were coaching their younger colleagues, transferring know-how and offering support in critical situations. After all, the coaching concept is not new to the Swiss Armed Forces. There is coaching for militia cadre as well as coaching courses for both militia and professional cadre. So far, however, a comprehensive coaching concept is still missing (Annen, 2006).

In the face of the issues mentioned so far, a *project proposal* was outlined and presented to the Secretary General of the Department of Defence by the summer of 2007. The most pertinent aspects of its content are listed below:

- There should be one coach per 25 career officers/NCOs. This coach should be a senior, experienced colleague, who has successfully completed a certified coaching course.
- Coaching is a offer free of charge.
- The coach is not a line manager, but rather perceived as a person of trust who can be consulted for specific needs in certain situations (e.g., work-life-balance, army-family adjustment, work technique, career issues, workplace conflicts, etc.). Also, the coach may act as an advisor to military leaders during a phase of either occupational or personal advancement.
- Mode, content, and goals of the coaching are agreed upon individually in a private conversation.
- Rather than just conveying general principles and messages in a detached manner, coaching offers the possibility of addressing leadership issues in a much more personal way. Moreover, it can serve as a seismograph monitoring the mental state of a military professional by having access to a continuous news flow.

- The project shall be monitored scientifically through a regular, systematic assessment of its effectiveness, while quality is assured via supervision and intervision.

This vision of "coaching for military personnel" and the associated ideas and objectives described above found an overwhelmingly positive response. However, it was made clear to us that though we were free to implement such a project, no resources could be allocated. Fortunately, by spring 2008 another staff programme got cancelled and the remaining funds were transferred to the coaching project.

The potential coachee's viewpoint

Prior to recruiting and training potential coaches, the demand for such an endeavor had to be appraised. Therefore, an inquiry was carried out on the basis of a full survey. All military professional cadre (N = 2111) were contacted via e-mail and asked to complete a questionnaire. A total of 865 persons responded, corresponding to a satisfactory return rate of 41%. The essential conclusions are summarized below:

- 68% of all respondents had experienced situations in which they would have appreciated the support of a coach; 18% indicated that these situations were frequent or very frequent.
- 33% were interested or even highly interested in a coaching programme conducted by an experienced colleague. A further 46% deemed it probable or highly probable to engage a coach.
- 64% of all respondents considered the benefit of a coaching programme to be high or very high.
- The idea of completing a coaching course and assisting colleagues as a personal coach appealed to 47% of all respondents.
- Generally, younger military professionals revealed a more positive and open attitude towards the coaching subject.

Further questions referred to the amount of time that military cadre were willing to dedicate to either a coaching programme or a coaching course. Last but not least, it is worth mentioning that most of the respondents commented on the requirements a coach would have to fulfill as well as on the possible situations in which they would enroll in a coaching programme. All in all it can be stated that the introduction of a "coaching for military personnel" was received favorably by a sufficient number of military professionals.

Given the fact that almost 60% of all addressees failed to respond and there were a number of critical and deprecatory voices even among the respondents, it was concluded that the project should be implemented in stages. That is, in a first stage a total of 10 coaches should be selected and trained. Only after the analysis and evaluation of this stage the next course of action should be defined. The latter could either imply the training and deployment of further coaches, a reduction or even the cancellation of the concerned activities altogether. The corresponding implementation concept was approved by the Armed Forces Command in the summer of 2008.

Selection of potential coaches

The recruiting process of potential coaches had to account for specific army-related aspects such as the equal consideration of all organizational units and language regions. On an individual level it had to ensure that candidates possessed enough professional experience to be perceived as credible by the coachees. Also, they would need to be available as coaches to the extent of their future assignment. Ultimately, eight career officers and eleven professional NCOs were designated after a prior consultation with their respective training units, thus yielding a total of 19 candidates. Each of them had to undergo a one-on-one assessment centre (AC) during half a day. This particular process had already proved to be of value for the selection of coaches for militia personnel. It encompassed the following elements:

- *Live Coaching:* The candidates have to demonstrate their coaching abilities within a real case scenario. Students of the military academy serve as coachees. They address personal experiences and problems that they have encountered while serving as militia company commanders or career officer aspirants. During the 50 minute coaching interview, the candidate is monitored by two experts, who appraise his aptitude in terms of presence, active listening, and systematic query, exploration, attitude and resource orientation.
- *Self/Third-party assessment:* Likewise, both coach and coachee assess the live coaching based on the above-mentioned criteria immediately after the live coaching session. They use, however, a shortened scoring sheet, which addresses only specific behavioral characteristics.
- *Structured interview:* The candidate is interviewed by two assessors, with the focus on his coaching competence. Questions touch upon the criteria the candidate uses to judge his own coaching activities and the measures he applies to assess his pursuit of improving his performance. Furthermore, particular situations are portrayed that could emerge in the course of a coaching session or during an entire coaching sequence. Relevant competences such as the

ability of reflection, flexibility, empathy or confrontation capability are assessed by analyzing and interpreting the candidate's behavior and statements.
- *Personality test:* Finally, the candidate completes the Bochum Business-focused Inventory of Personality (BIP). Based on previous experience with this instrument the relevant dimensions for coaching activities are well known.

The respective ACs took place within a time span of ten days, each immediately followed by a subsequent first appraisal. After completion of all ACs and reviewing the performance of each candidate, it was decided whether or not the respective person should be admitted to the coaching course based on the evaluation. For twelve candidates the answer was a clear "Yes", for six it was a clear "No", while in one inconclusive case the candidate was ultimately rejected. As a result, a total of twelve military professionals (consisting of five career officers and seven career NCOs) were notified in a personal interview that they were considered eligible as coaches. Likewise, the negative decision was communicated to the remaining seven candidates in a private conversation.

Training and deployment

Meanwhile, the instruction contents and potential providers of customized coaching education were evaluated. The primary requirement for such an institution was its readiness to cooperate with a military organization and its distinct culture and constraints, notably a relatively limited number of mandatory contact hours. A second precondition was the willingness of the counterparty to engage in a long-term collaboration. It was therefore not surprising that the Zurich University of Applied Science (ZHAW) won the tender against two consulting firms. After all, the Institute for Applied Sciences (IAP) at the ZHAW has a long track record in the field of instruction and training of coaches and supervisors. It enjoys a correspondingly sound reputation and disposes of all the human resources necessary.

Experts of the IAP and the project team "coaching for military personnel" cooperatively compiled a 12-day training programme, consisting of six instruction modules to be attended between April 2009 and January 2010. Successful completion would grant a Certificate of Advanced Studies in Coaching (CAS Coaching), and the programme addressed the following topics in a hands-on manner:
- Clarification of a coach's status and duties
- Attitude and apperception processes in counseling
- Stages of coaching

- Testing and exercising various interrogative forms and counseling interventions
- Coaching in conflict situations
- Coaching as an element of transformation processes
- Conduction of case studies

From the start of the second instruction module, the coaches should already work in practice. Consequently, their *coaching profile* is visible on the intranet page for military staff. It includes contact details, information on current activities and professional experience, the individual coaching motto, as well as details on the possible assignments and coaching languages. The reason for the early practical engagement is threefold: first, all candidates have a sufficiently profound professional background and experience of life. Second, they can resort to a basic knowledge in the field of coaching. Third, this approach allows for the examination of tangible and current cases in the training and group supervision process.

Evaluation

It was clear from the very beginning that the project "coaching for military personnel" would be monitored scientifically. One reason is the onus of providing the armed forces with solid information on the effectiveness of coaching. Another is that contributing to a scientific discourse in a popular, yet sparsely explored domain is always a worthwhile opportunity.

Therefore, while planning the education and practical assignments, an evaluation procedure was developed that incorporated both the specific project circumstances and the relevant literature available on this subject (e.g., Greif, 2007; Jansen, Mäthner & Bachmann, 2003; Künzli, 2005). In an iterative intercommunication process involving all project team members, an evaluation tool was developed in order to gather important information prior to and after each counseling sequence. Thereby both coach and coachee were given the opportunity to comment on the relevant questions.

In accordance with the literature pertinent to the topic, the scientific evaluation focused on three main areas: *Structure – Process – Outcome*. In what follows, its substance is illustrated by means of concrete items:

- In the *structure* section, the characteristics of both coach and coachee are recorded ("My coach helped me to find my own solutions"; "The coachee was willing to change certain things"; etc.). Another set of questions refers to the relationship between coach and coachee ("The relationship was marked by mutual respect"; "… by confidence"; etc.). Lastly, procedural factors are

addressed such as location, time, and duration ("The room was suitable for a coaching session"; "Appointments could easily be arranged"; etc.).
- The *process* section concerns the crucial aspects of contract clarification, goal definition, transparency, and participation. Also in this section, the general perception of the coach is addressed. So are questions about his professional skills ("My coach applied methods appropriate for this particular situation"; "… together with me he developed concrete measures"; etc.) as well as questions concerning the role assumed by the coach and the way he was perceived by the coachee (e.g., intimate, professional, challenger, role model, etc.).
- The *outcome* section includes an evaluation of the overall changes ("Thanks to the coaching I have gained more clarity on my situation"; etc.). In addition, the persons involved should account for the achievement of objectives. Also, they must appraise the cognitive, emotional and behavior-related impact ("Through the coaching I have become acquainted with more behavioral patterns"; "I have improved my self-esteem"; etc.). The section concludes with a general assessment of the entire coaching sequence ("I am very pleased with the outcome of the coaching"; "I would recommend my coach to others"; etc.).

After the content of the questionnaire was determined, it was released online. This enables the project management to send a web link to the coach and coachee, prompting them to answer the respective questions in a timely manner, both after the initial contact and after conclusion of the coaching. This procedure ensures a constant news flow from current and recent coaching sequences.

Generally it can be stated that a tailored evaluation tool has been developed based on a solid foundation of theoretical considerations. It systematically considers practical experiences and should provide a starting point for the systematic reflection and improvement of the procedure. From the requester's and project management's viewpoint both controlling and quality control are warranted. Moreover, the gained insights can be incorporated in the optimization process of the specific training. From a scientific perspective, significant contributions to the discussion on coaching effectiveness should be able to be made in the long term.

Current status and outlook

When recapitulating the project "coaching for military personnel" and its characteristic elements, the current project status is strikingly and pleasantly close to the initially mentioned proposal. The only notable exception is the mismatch between the expected and effective number of available coaches. However, in light of the required resources it would not be advisable to offer a full coverage already at this early stage.

With regard to content, however, the objectives are implemented with the provision of a professional and practical assistance free of charge. Though not a line supervisor, the coach is familiar with most operation-specific questions. Hence, the coachee can address problems without having to fear any negative impact on staff reports or promotion decisions later on. The proposition was made known to all professional military via bulletin. The profiles of all coaches are accessible on the intranet and a personal contact can be established in a straightforward and unbureaucratic manner.

Nevertheless, after the first two years of the "coaching for military personnel" project the observed demand is still moderate. Given the specific organizational culture of the military environment, this is of little surprise, however. A number of analyses have shown military personnel to be particularly hesitant in accepting psychological support (e.g., Hoge, Castro, Messer, McGurk, Cotting & Koffman, 2004). The anxiety to admit personal shortcomings is simply too predominant, as is the fear of losing the confidence of one's comrades and fellows. Accordingly, there is a desire to arrange counseling interviews incognito and off the job. This in turn entails organizational challenges.

As elaborated above, "coaching for military personnel" accounts for such concerns. Overcoming general misconceptions, however, remains essential. It is conceivable that word-of-mouth advertising will prove the most effective means to overcome such mental barriers. Basically, the chosen strategy will be pursued, while first and foremost the emphasis will be on the quality of the coaching sessions initiated so far.

At the same time, an exchange of information is maintained both with corporations and scientific institutions dealing with the subject of coaching for practical and theoretical reasons. Clearly, the cooperation in a military context is of pre-eminent interest. Specific coaching projects have been launched by the Singapore Armed Forces (Ramaya, Wee, Chan & Lim, 2006) and by the German Bundeswehr (Kaufel, Scherer, Scherm & Sauer, 2006). In the Austrian Bundesheer, conceptual preparations are already at an advanced stage (Dengg, 2006). The cooperation mentioned should be integrated, since the clash of opinions spawns reality. Thus, coaching in military organizations will be given the position it deserves in light of the prevailing challenges faced by military personnel.

Bibliography

Annen, H. & Eggimann, N. (2006). *Assessment Center result as predictor for study success.* Paper presented at the 48th Annual Conference of the International Military Testing Association (IMTA), Kingston/CA, October 2-5, 2006.

Annen, H. (2004). *Personnel Marketing in Turbulent Times. Recruitment and Retention Activities for Military Personnel in the Swiss Army.* Paper presented at the 46th Annual Conference of the International Military Testing Association (IMTA), Brussels, October 26-28, 2004.

Annen, H. (2006). Coaching – alter Wein in neuen Schläuchen? In H. Annen & U. Zwygart (Eds.), *Das Ruder in der Hand. Aspekte der Führung und Ausbildung in Armee, Wirtschaft und Politik* (pp. 127-134). Frauenfeld; Stuttgart; Wien: Huber.

Annen, H. (2007). Leadership as a Selection Criterion for Officers – The Assessment Center for Prospective Career Officers (ACABO) in the Swiss Armed Forces. In H. Annen & W. Royl (Eds.), *Military Pedagogy in Progress* (pp. 195-212). Frankfurt a.M., Berlin, Bern; Bruxelles, New York, Oxford, Wien: Peter Lang.

Bergmann, B. (2000). Arbeitsimmanente Kompetenzentwicklung. In B. Bergmann, A. Fritsch, P. Göpfert, F. Richter, B. Wardanjan & S. Wilczek (Eds.), *Kompetenzentwicklung und Berufsarbeit* (pp. 11-40). Münster; New York: Waxmann.

Dengg, O. (Ed.) (2006). *Coaching. Ein Instrument für Management und Führung.* Wien: Schriftenreihe der Landesverteidigungsakademie.

Greif, S. (2007). Advances in Research on Coaching Outcomes. *International Coaching Psychology Review, 2* (3), 222 – 249.

Gutknecht, S. & Krautz, A. (2004). Organisationsspezifische Einstellungen Schweizer Berufsmilitärs. *Allgemeine Schweizerische Militärzeitschrift, 170* (3), 24-26.

Gutknecht, S. (2001). *Arbeitszufriedenheit und Commitment in Zeiten organisationalen Wandels. Zum Einfluss von Persönlichkeitsmerkmalen auf organisationsspezifische Einstellungen – eine Untersuchung in Militär und Wirtschaft.* Unpublished PhD thesis, University of Zurich, Faculty of Philosophy.

Herzberg, F. (1974). *Work and the nature of man.* London: Staples Press.

Hoge, C., Castro, C., Messer, S., McGurk, D., Cotting, D. & Koffman, R. (2004). Combat Duty in Iraq and Afghanistan, Mental Health Problems, and Barriers to Care. *New England Journal of Medicine, 351*, 13-22.

Jansen, A., Mäthner, E. & Bachmann, T. (2003). Evaluation von Coaching. Eine Befragung von Coachs und Klienten zur Wirksamkeit von Coaching. *Organisationsberatung – Supervision – Coaching, 3*, 245-254.

Kaufel, S., Scherer, S., Scherm, M. & Sauer, W. (2006). Führungsbegleitung in der Bundeswehr – Coaching für militärische Führungskräfte. In W. Backhausen & J.-P. Thommen (Eds.), *Coaching. Durch systemisches Denken zu innovativer Personalentwicklung* (pp. 419-438). Wiesbaden: Gabler.

Künzli, H. (2005). Wirksamkeitsforschung im Führungskräfte-Coaching. *Organisationsberatung – Supervision – Coaching, 3*, 231-243.

Ramaya, R., Wee, E., Chan, K.-Y. & Lim, K. (2006). *Measuring Effective Coaching in the Singapore Armed Forces.* Paper presented at the 48th Annual Conference of the International Military Testing Association (IMTA), Kingston/CA, October 2-5, 2006.

Semmer, N. , Tschan, F., Meier, L., Facchin, S. & Jacobshagen, N. (2010). Illegitimate tasks and counterproductive work behavior. *Applied Psychology: An International Review, 59,* 70-96.

Stocker, D., Jacobshagen, N., Semmer, N. & Annen, H. (2010). Appreciation at Work in the Swiss Armed Forces. *Swiss Journal of Psychology, 69* (2), 117-124.

Luiza Kraft

Enhancing Action Competence in Military Higher Education through Cooperative Learning

Introduction

In these days of globalization, intercultural communication plays an important part in English Language Teaching for several reasons, one of them being that English has assumed the ever-increasing role of an international language used extensively by millions of people outside its original geographic boundaries, to "convey national and international perceptions of reality which may be quite different from those of English speaking cultures" (Alptekin, 1984, p. 17). As English continues to spread as an international language, the number of second language users of English will continue to grow, thus outnumbering the native speakers of English by far. The Romanian military has its stake in this as well, being an integral part of NATO. Within the multinational forces that make up NATO and its PFP partners, Romanian military personnel use English more often to communicate with comrades for whom English is a foreign language, too, than with those for whom it is their native tongue. It is apparent that "English is the main link language across cultures today" (Schnitzer, 1995, p. 227). Therefore, the goal of learning English shifts from enabling learners to communicate their ideas and culture not only with native Anglophones, but also with those of other cultures. Consequently, the question of intercultural communication is clearly indispensable in English language learning and teaching if the aim is to develop students' communicative competence, as well as their action competence in general.

Cooperative learning in the military higher education

Maybe there is no other group in which teamwork in real-life situations is more important than the military. Let us take for instance the example of task forces – *combined* as in CJTF[1] or *deployable* as in DJTF[2], fighter pilots engaged in dog-

---

1   Combined Joint Task Force: a multinational task force that can be used rapidly as a peacekeeping force.
2   Deployable Joint Task Force: a task force that can be rapidly moved to a war zone or area of operations.

fights, or Special Forces, where the survivability of the group depends on individual performance and vice versa. The adjective *joint* has acquired a special meaning in the military domain, by designating two or more services operating together. The idea of 'togetherness' is of utmost importance in current military collocations, such as *joint warfare, joint operations, combined/deployable joint task force, joint decision-making, joint planning,* etc. pp.

In addition to the above, NATO's Article 5 of the Washington Treaty states that an attack upon one member of the Alliance is an attack upon all its members. Moreover, in 1994 the *Partnership for Peace (PfP)* was set up beyond the NATO geographical borders, both as a concept and as a practical ground in order to prepare the accession of other European countries to the Alliance. We may therefore conclude that the philosophy of unity in diversity, togetherness, partnership, team work, and cooperation prevails in everything the servicemen and women think, carry out, or plan to do.

But in order to be able to work and cooperate in teams, training methods that recreate in the classroom environment an atmosphere of cooperation, sharing and mutual learning need to be used. The task is difficult, the more so as we are talking about an organization based on hierarchy, chain of command, and a very rigid distribution of roles.

The main outcomes of collaborative learning methods used in the military higher education, as far as English language courses are concerned (Kraft & Tutuianu, 2008), have proven successful for endowing military students with the ability to:

- use teamwork as a guiding principle for both classroom and job-related tasks;
- lead and take part in a discussion using various language functions in order to hypothesize, support opinion, agree and disagree, etc.
- reach consensus;
- jointly choose the best course of action;
- adequately interpret cultural resemblances and differences and mediate cultural misunderstandings;
- build a successful multinational team;
- develop interpersonal skills by communicating and sharing knowledge, beliefs and experience;
- become more flexible and tolerant.

The general profile of the military students

Our students are military personnel, ranking from junior lieutenants to colonels and sometimes higher, participating in peace support operations in various areas of the world, carrying out missions ranging from conflict prevention and crisis response to peace enforcement, peacekeeping, and humanitarian aid. In peacetime they may work in multinational regional or North Atlantic Treaty Organisation (NATO) commands, whereas in times of crisis they are deployed to the theatres of operations where they support the peace process under UN mandate. This two-fold role requires linguistic competence in a foreign language in order to communicate and process information and, more importantly, the ability and willingness to develop a human relationship with people of other languages and cultures.

Description of the project

The project I want to present was designed to be implemented in one intensive, pre-deployment course (70 hours) of PSO English, delivered for two similar groups, as follows:

Course language:
English
Student criteria:
Officers ranking from Second Lieutenant to Colonel
Language competence:
2222 or 3232 measured on the STANAG – 6001, or B2 according to CEF standards (Council of Europe, 2001).
Students from other countries:
NATO and PfP member states (up to four foreign students per group)
Note: In the sample group (further referred to as Group 1) there were 8 Romanians, 1 Croat, 1 Azerbaijani, 1 Greek and 1 Moroccan.
In the control group (or Group 2), there were 8 Romanians, 1 Croat, 2 Bulgarians and 1 Georgian.

For Group 2 the traditional student-centered approach was used, while for Group 1 a different teaching and learning principle was tried, in which students were put to work together to accomplish shared goals; more explicitly, students were trained to work together to maximize their own and one another's learning.

The goal of the project was to identify and experiment with language teaching methods that can contribute to greater mutual understanding and acceptance of other perspectives, can build up character by building group confidence and cohesion, train people to work in teams, and share knowledge. We also sought to empower the students and transfer to them the responsibility for the achievement of the course objectives and the success of overall and separate activities.

With this in mind, we needed a field-tested teaching strategy for small groups that could improve our students' (ability ranging from B1-B2 to C1 on the CEF) understanding of a subject, in our case both general and military topics. The syllabus was the generally accepted one for this type of course. Within this project, each member of a team was made responsible not only for learning what was taught but also for helping team-mates learn, thus creating an atmosphere of achievement. Students were supposed to work through the assignment until all group members had successfully understood and completed it. The method (or rather process) we chose was *cooperative learning (CL)*.

Cooperative efforts were expected to help enhance the participants' efforts towards mutual benefit so that all group members would:
- gain from one another's efforts;
- recognize that all group members share a common fate;
- know that one's performance is the mutual result of oneself and one's team members;
- feel proud and jointly celebrate when a group member is recognized for achievement.

Research made during and after the project development has shown that by using cooperative learning techniques it was possible to:
- enhance student satisfaction with their learning experience;
- help students develop skills in oral communication;
- develop students' social skills;
- promote student self-esteem;
- increase self-confidence in using English;
- help promote intercultural understanding.

During the project it became obvious that cooperative efforts may be expected to be more productive than competitive and individualistic efforts, but only under certain conditions. These conditions (see Johnson, Johnson & Holubec, 1993; Johnson & Johnson, 2001, Kagan, 2001), once explained and implemented, commented on by both teachers and students, and internalized, were as follows:

1. *Positive Interdependence* (i.e., sink or swim together)
- Each group member's efforts are required and indispensable for group success;
- Each group member makes a unique contribution to the joint effort because of his/her resources and/or role and task responsibilities.

2. *Promotive Interaction* (i.e., promoting one another's success)
   - Orally explaining how to solve problems;
   - Teaching one's knowledge to others;
   - Checking for understanding;
   - Discussing concepts being learned;
   - Connecting present with past learning.

   Each of these activities can be structured into group task directions and procedures, for it is through promoting one another's face-to-face learning that members become personally committed to one another as well as to their mutual goals.

3. *Individual & Group Accountability* (i.e., the group is accountable for achieving its goals and each member is accountable for contributing his/her share of the work):
   - Working with small groups to provide for greater individual accountability. The class consisted of 12 students, usually divided into three groups of four;
   - Periodically giving an individual progress test to each student;
   - Randomly examining students orally by calling on one student to present his/her group's work to the teacher (in the presence of the group) or to the entire class.
   - Observing each group and recording the frequency with which each member contributes to the group's work;
   - Assigning the role of mediator, group discussion leader or assessor to one student in each group. The assessor's task was to ask other group members to explain the reasoning and rationale underlying the group's answers;
   - Having students teach what they learned to someone else.

   Practice has shown that individual accountability exists when the performance of each individual is assessed and the results are communicated to the group and the individual in order to ascertain who needs more assistance, support, and encouragement in learning. By making students learn together, they subsequently gained greater individual competency.

4. *Interpersonal & Small-Group Skills*
   - Leadership
   - Decision-making
   - Confidence-building
   - Communication
   - Conflict-management skills

   Procedures and strategies for teaching students social skills may be found in Johnson (1991, 1993) and Johnson and Johnson (1988).

5. Group Processing
   - Have group members discuss and describe how well they are achieving their goals and maintaining effective working relationships;
   - Identify which member actions are helpful or not;
   - Make decisions about what behaviors to maintain or change.

Throughout the course we could notice that careful analysis of how members are working together and determining how group effectiveness can be enhanced resulted in continuous improvement of the learning process.

The implementation of CL methods required from the project designers to implement changes on the lesson level so that students could in fact work cooperatively with each other. What we actually had in mind was to:

- take the existing topics and course contents and structure them cooperatively;
- tailor CL lessons to meet the military instructional circumstances and needs of the course type, syllabus, subject areas, and students' profile;
- foresee and be able to diagnose the problems students might encounter in working together;
- intervene whenever and wherever necessary in order to increase the effectiveness of the learning groups.

The activities included in the project addressed the basic skills, including reading materials, taking notes, listening to each other, delivering presentations, and discussing/debating. There were some pre-set groups, to ensure an equal balance of language ability, nationality, age, and service branch. At some points the groups were interchanged, so that in the end all the students had gotten the opportunity to work closely together with as many other students as possible, thus experiencing as many different working styles, perspectives, Englishes, behaviors, and human reactions as possible.

On the whole, the goals of the cooperative activities were of two kinds:

1. Language-related:
   - Developing students' sub-skill of understanding spoken and/or written materials on concrete or abstract, familiar or non-familiar, job-related or general complex topics;
   - Developing students' sub-skill of delivering a speech or presentation on a complex topic, integrating pertinent examples, secondary arguments, and specific points so as to reach an adequate conclusion;
   - Developing students' capacity of leading and taking part in a group discussion using various language functions in order to hypothesize, support opinion, agree and disagree, etc., so as to reach an agreement and build consensus.

2. Intercultural education-related:
   - Developing students' capacity of adequately interpreting cultural resemblances and differences and mediating cultural misunderstandings;
   - Increasing their effectiveness in working as a group by developing interpersonal skills, sharing beliefs and experience and becoming more flexible.

The necessity of intercultural education in the 21st century armed forces

The prime objective of embedding an 'intercultural dimension' in military foreign language teaching is to train learners as *intercultural speakers* or *mediators*. When designing the project, we started from the premise that in order to successfully accomplish peace support operations, our students would need to be militarily prepared, but they should also be able to face the challenges of multiculturalism, i.e., to engage with complexity and multiple identities and to avoid the stereotyping which occurs when perceiving someone through a single identity. The intercultural dimension of foreign language learning is based the acquisition of skills complementary to linguistic competence which allow the learner to perceive the interlocutor as an individual whose qualities are to be discovered, rather than as a representative of an externally ascribed identity. In addition, intercultural communication consists of respect for individuals and equality of human rights as the democratic basis for social interaction, an approach discussed by Byram (2002). Accordingly, language teaching with an intercultural dimension helps learners to acquire the *linguistic competence* needed to communicate in speaking or writing, i.e., to formulate what they want to say and write in correct and appropriate ways. But it also develops their *intercultural competence*, i.e., the ability to ensure a shared understanding by people of different cultures and to interact with people as complex human beings with multiple identities and their own individuality.

Therefore, the intercultural dimension of foreign language teaching is embedded in the Romanian military higher education through the following learning objectives:

- to give students intercultural competence as well as linguistic competence in the target language;
- to prepare them for interaction with people of other cultures;
- to enable them to understand and accept people from any other culture as individuals with other perspectives, values, beliefs, and behaviors;

- to make them tolerant of 'otherness' in order to avoid passing judgments in this respect;
- to help them perceive the exposure to another culture and the interaction with it as a positive and enriching experience.

The intercultural dimension of foreign language courses

In order for the linguistic programmes conducted in the Romanian Ministry of Defence to successfully meet their objectives and commitments under the Common European Framework for Reference (2001) provisions, i.e., to train not only foreign language speakers, but first and foremost intercultural speakers/mediators, the following practical methods have been developed in our institution beginning with 2005 (Kraft, 2007):

- Inviting native speakers of the target language(s) to participate in the students' debates focusing on intercultural topics, in order to facilitate the expression of different perspectives, as well as to answer the students' questions;
- Creating opportunities for officers or military instructors who have participated in missions on the theatres of operations in the Middle East or Africa to share their own experiences and to highlight the cross-cultural impact in those areas, both during and post-conflict;
- Including syllabuses in the foreign language course activities which might help raise the awareness of both one's own and the target culture;
- Asking the foreign students at "Carol I" National Defence University to take part in the debates on intercultural topics and encouraging them to express their points of view, with two precise objectives in mind:
  - to facilitate the information exchange on traditions, values, and attitudes in their own culture, and
  - to facilitate their Romanian fellow students' understanding of the way in which foreigners perceive our traditions, culture, and values.
- Including students' research topics on intercultural issues in all the foreign language syllabuses;[3]
- Organizing group discussions & seminars on intercultural education, with the aim of facilitating students' perception of it as a solution to certain current issues, e.g., interethnic violence, immigration and demographic change, historical conflicts, prejudices, etc.;

---

3   These research topics are materialised into briefings or end-of-course projects which the students have to prepare as part of the intensive or extensive linguistic programmes, irrespective of the level of training (i.e., graduate or postgraduate).

- Solving specific problems by applicative exercises and case studies in order to develop students' cognitive capabilities with respect to other cultures, as a response to the challenges of a cultural environment different from the Romanian one. To this end the following topics have been used:
  - An analysis of models of social cohesion, integration, and problem solution within some contemporary multicultural societies, such as the USA, Canada, Australia, Switzerland, Belgium, etc.;
  - A conflict analysis of societies divided on the basis of culture;
  - Cultural relativism and reciprocity of influences;
  - Different cultures' perception of: time and space; gender roles; human relationships; family relationships; humor; non-verbal communication, etc.
  - Principles of education in different cultures, etc.

Let us therefore keep in mind that in order to develop the intercultural openness of representatives of the military institution, the role of any foreign language teacher is to develop attitudes and value awareness in their students' skills, complementing the language acquisition for both short and long term purposes.

Results and conclusions of the project

The analysis of the project results has shown that students in sample Group 1, who had opportunities to work collaboratively:

- learned faster and more efficiently,
- met the course objectives to a greater extent,
- exhibited greater retention,
- felt more positive about the learning experience than control Group 2.

The course critique for Group 1 clearly and unanimously stated that the CL methods used to deliver information and create knowledge helped students learn essential interpersonal skills and develop the ability to work collaboratively – a skill now in great demand not only in the national military workplace, but also in the NATO multinational environment. Students rotated different roles – depending on the activity – such as briefer, teacher, facilitator, group discussion leader/participant, mediator, and monitor. During the activities designed or adapted for the cooperative group, every participant had a specific task, tailored to their competencies, so that everyone could be involved in the learning or project and contribute to the overall success of the group.

According to the end-of-course questionnaire, we can classify the outcomes into three major categories: achievement of the course objectives, positive relationships, and excellent rapport. The analysis clearly indicated that cooperation, compared to competitive and individualistic efforts, resulted in higher personal achievement, more supportive and committed relationships, and greater social competence and self-esteem.

Moreover, Group 1 students were more positive and caring about each other, regardless of differences in their linguistic competence, ethnic backgrounds, experience, or education. From an interpersonal viewpoint, they were also more effective, in the sense that they gradually became better prepared to assume the perspectives of others. They also developed more positive attitudes about taking part in controversy or conflict situations, had developed better interaction skills, and as opposed to their peers in Group 2 who learned in a competitive and individualistic setting, had a more positive expectation about working with others.

As to the students of Group 2, they worked mostly against each other to achieve an individual aim. As in any competition, there was negative interdependence among goal achievements, insofar as students thought they could meet their goals only if the other students in the class failed to meet theirs (Deutsch, 1962; Johnson & Johnson, 1988). Consequently we could notice that Group 2 students either worked hard to do better than their classmates, or they took it easy because they did not believe in their chance to succeed on their own. All in all, Group 2 students worked alone to accomplish goals unrelated to those of their group members, clearly resulting in a focus on self-interest and personal success, while the achievements and failures of others became irrelevant.

Nevertheless, as in any human activity, there is always room for improvement. Even seemingly positive innovations can have unintended and negative consequences and students may resist collaboration even while simultaneously recognizing its value.

In the course feedback, there has been some criticism as to the method used, and some (but very few) of the students were not very satisfied with certain aspects of cooperative learning. These might be classified into three major categories, as being related to:

1. student involvement and sharing responsibilities;
2. teachers;
3. assessment.

As far as the first point is concerned, there were opinions which revealed dissatisfaction with the method, because they felt that in each group some people carried out the tasks and made the group successful, while others leant back comfortably and indulged in being bumbled through most of the activities.

Some of the flaws have been diagnosed as originating from the fact that certain people are reluctant to share and cooperate as far as learning and especially assessment activities are concerned; consequently, the issue here is to enhance socialization at an early age with a focus on fostering team work and cohesion-building.

In addition, the project results have proven that in order to get interactive classrooms, more needs to be done in terms of teacher training. Let us remember the words of Fred Korthagen (2006): "Consciously, teachers teach what they know. Unconsciously, they teach who they are". It is therefore the teacher's role to observe students, to create equal opportunities for expression, to give clear and unambiguous instructions, examples, and demos, and last but not least, to produce and use objective assessment criteria in an environment where 'think-pair-share' is the guiding principle. These skills need to be taught to teachers, and more emphasis needs to be placed on this issue in the teacher training seminars and workshops.

Another critical remark is that the effects of cooperative learning were not automatic. Partnerships in the classroom took some time to develop, and the course was quite too short to harvest the full crop.

The assessment-related criticism is closely connected to the aforementioned points. Some students were surprised that they were also assessed collectively, as they considered that certain people got high scores because others did the work.

Enhancing group effectiveness both professionally and language wise

At the end of the activities participants were asked to think back to their work as a team and discuss what they found useful, challenging, interesting, etc., in working together. After that, the teacher carried out a debriefing on how the group had functioned. Encouraging reflection on their experiences, the questions asked centered around the learning experience the teacher had wanted the students to achieve. Some examples of questions were:

- How did you reach decisions?
- What role did each member adopt?
- Did you listen to each other?
- Does the preliminary project resemble the final proposal?
- How did you manage to come to a common decision although in the beginning you had different ideas with respect to, e.g., the most important qualities of a leader?
- What did you learn about the functions of a group?
- What would you do next time in order to improve joint decision-making?
- What made your group work and what made you successful as a group?

The answers were more or less the same for all groups, and by confronting the entire class with them, the students became once more aware of what makes a group successful. The key issues pointed out were as follows.

In order to be effective as a group both professionally and languagewise, participants have to:

- feel close to and responsible for each other and the accomplishment of the group tasks;
- build a supportive, positive atmosphere;
- listen to and learn from one another;
- communicate openly;
- compromise and make decisions together despite their differences;
- be open to learn about one another's assumptions and beliefs (most of which are cultural);
- understand, trust, and accept each other, which means *to cooperate*;
- confront possible differences of opinion openly;
- settle contradictions constructively;
- have a sense of humor.

At some point during the project, the students were asked to write down in groups a set of 'rules of engagement' in a military group discussion. Having compared and put together what the three groups came up with, this is a summary of their recommendations:

1. A discussion leader or moderator must be completely objective about the facts.
2. He/she should never give personal opinions.
3. A discussion leader must not permit a few members of the group to dominate the discussion, but should encourage everyone to participate.
4. The moderator must insist that every participant states his/her idea on the subject. Occasionally, the moderator must direct a question to a person who has been silent for too long.
5. He/she must not let the discussion become one-sided, and invite opposing ideas into the discussion.
6. The moderator must guide the participants towards a conclusion/resolution of the problem.
7. Individual participants must always be supportive of the group.
8. Participants must be respectful towards the opinions of their colleagues.
9. Participants must view the same facts and circumstances from different perspectives and have an open mind.
10. Participants must exchange helpful ideas and information relevant to the discussion or problem.

11. There must be no room for competition among participants, only for improvement of language and interpersonal skills.
12. Participants must be patient and careful about interrupting.
13. Participants must remember there is no right or wrong solution, but only the best solution for the group.
14. All participants must work as a team.

The way ahead

The project on cooperative learning for intercultural education has resulted in positive aspects and outcomes. Consequently, one of the on-going tasks of the people responsible for the project is to translate the concept of cooperation into a set of practical strategies for use by the foreign language teachers in "Carol I" National Defence University. On a short-term perspective, we are planning to organize a set of methodology workshops meant to train the teaching staff on the strategies of structuring cooperative interactions and teaching students the skills needed to work effectively with others (i.e., communication, leadership, confidence-building, conflict resolution). A basic model has been adapted to meet military foreign language training needs, focused on what decisions the teacher needs to make before a lesson, what to tell the students at the beginning of the lesson to activate the cooperative goal structure, and what role the teacher takes on as the students are working. An outline of the model should include, among other aspects, the following (see Johnson, Johnson & Holubec, 1993):

1. Selecting a type of lesson and adapting its contents. Our cooperative learning groups have shown to be especially effective where problem-solving and joint decision making are involved.
2. Selecting the group composition. We could notice once more that heterogeneous groups tended to be more powerful than the homogenous ones. One explanation would be that by learning in cooperative groups the need for discussion, explanation, justification, and shared resolution on the subject matter arises. Quick consensus without any argument does not enhance learning as effectively as does the discussion of different perspectives, considering different alternatives, supporting members who need help, etc.
3. Providing the appropriate materials and appropriate techniques, e.g., "jigsawing" the material and allocating different responsibilities within the group so that all students have duties associated with their piece of the assignment (i.e., reading to the group, researching and reporting back for discussion, explaining and offering support, etc).

In conclusion, the implementation of the project has demonstrated that students working together cooperatively progress faster. The overall learning process not only has positive effects on the classroom climate, but also teaches the students to accept differences, to become more tolerant and to transform negative attitudes into positive ones. Being able to perform language skills such as listening, speaking, reading, or writing, is valuable indeed. But such skills are of little use if the person cannot apply them in a cooperative interaction with other people in professional and/or multinational settings.

Bibliography

Alptekin, C. (1984). The question of culture: EFL teaching in non-English-speaking countries. *ELT Journal, 38* (1), 14-20.

Byram, M., Gribkova, B. & Starkey, H. (2002). *Developing the Intercultural Dimension in Language Teaching. A Practical Introduction for Teachers.* Language Policy Division. Directorate of School, Out-of-School and Higher Education, DGIV. Strasbourg: Council of Europe.

Council of Europe (2001). *The Common European Framework for Reference for European Languages: Learning, Teaching, Assessment.* Cambridge: University Press.

Deutsch, M. (1962). Cooperation and trust: Some theoretical notes. In M. R. Jones (Ed.), *Nebraska Symposium on Motivation* (pp. 275-319). Lincoln: University of Nebraska Press.

Johnson, D. (1991). *Human Relations and Your Career* (3rd. ed.). Englewood Cliffs: Prentice-Hall.

Johnson, D. (1993). *Reaching out: Interpersonal Effectiveness and Self-Actualization* (6th ed.). Needham Heights: Allyn & Bacon.

Johnson, D. & Johnson, R. (2001). *Cooperative Learning.* Retrieved from http://www.clcrc.com/pages/cl.html

Johnson, D., Johnson, R., & Holubec, E. (1993). *Cooperation in the Classroom* (6th Ed.). Edina: Interaction Book Company.

Johnson, R. & Johnson, D. (1988). Cooperative Learning. Two heads learn better than one. *In Context, 18*, 34-36.

Kagan, S. (2001). *Structures for Emotional Intelligence.* Retrieved from http://www.cooperativelearning.com/free_articles/dr_spencer_kagan/ASK14.php.

Korthagen, F., Loughran, J. & Russell, T. (2006). Developing fundamental principles for teacher education programs and practices. *Teaching and Teacher Education, 22*(8), 1020-1041.

Kraft, L. (2007) *Foreign Language Learning and its Role in Developing Eurocompetences and Intercultural Communication Skills.* Military Journal of Management and Education, 1, 5-12.

Kraft, L. & Tutuianu, D. (2008). *Cooperative Learning for Intercultural Education in Foreign Language Learning.* Paper presented at the IAIE and IASCE International Conference for Cooperative Learning in Multicultural Societies: Critical Reflections, January 19th to 22nd in Turin, Italy.

Schnitzer, E. (1995). English as an international language: implications for the interculturalists and language educators. *International Journal of Intercultural Relations, 19* (2), 227-236.

Ulla Anttila

Military Pedagogy and Human Security Education

Introduction

The concept of human security was introduced in the 1990s to broaden the state-based security approach and take into account an individual's needs for security. Human security education of military personnel should be of interest in military pedagogy. As part of a soldier's ethical reasoning he or she needs the capacity to handle questions related to every individual's right to security and a safe environment. Regarding today's military personnel's tasks, information on human security and practical strategies on how to implement it is especially required in peacekeeping and crisis management activities, where contacts to local people are extremely important. Multiculturalism makes these contacts in post-conflict areas demanding.

In this article I introduce and analyze military pedagogy and human security conceptually. In the empirical part of the article, I analyze Finnish cadets' answers to a questionnaire focused on human security. In December 2007, cadets and officers in the fourth year of the Finnish National Defence University cadet course were asked to describe their thoughts on human security issues. They were asked how the concept of human security had emerged during their studies at the National Defence University, how they found the concept adapted to their future career as officers, and how they thought human security should influence the future development of international crisis management.

Military pedagogy and its challenges

A military-pedagogical framework works out holistic and critical views of soldiers and officers as learning, growing, and acting human beings who interact with their environments (Toiskallio, 2000a). It is a human science within military sciences (Toiskallio, 2000b). Military pedagogy should be seen as a multi-disciplined field of research: sociology, psychology, pedagogy, and studies related to sports have an impact on it (Toiskallio, 1996). Holistic interpretations about human beings should not exclude applying results from other fields of science into military pedagogy. These four disciplines mentioned by Toiskallio (ibid.) may broaden the traditional military-pedagogical framework significantly.

Pedagogy refers to both practice and science or theory. The practical aspect of military pedagogy is military education and training. The scientific aspect of military pedagogy is critical reflection of military education and training. In pedagogy both theory and practice belong together. Military training is a practical issue, while the tools of military pedagogy are research and teaching (Toiskallio, 2000b). The main purpose of military pedagogy as a practical science is to produce knowledge for, and make contributions to the military training doctrine of armed forces (Toiskallio, 2000b).

The term "military pedagogy" has been used in its contemporary meaning in Finland since 1996 (Toiskallio, 2000b). In Switzerland, military pedagogy is defined as "the science of education, training and leadership of human beings in a military environment. It deals with the individual as well as with interpersonal relations in the social fabric of the military" (Annen, 2000, p. 33). Schunk and Nielsson (2002) define the objective of military pedagogy as solving the problems connected with learning in relation to military education and training. In comparison to Toiskallio (2002), their perspective on the tasks of military pedagogy is narrower.

Because of the demanding nature of the future military duties requiring a multitude of skills, it is wise to maintain the holistic perspective on the individual as a key principle in military pedagogy. Complex conflicts and the development of military crisis management challenge military pedagogy. It is substantial to carry out more empirical research on the education and training required in crisis management.

Peace operations and military crisis management have seldom been a subject of analysis in terms of learning environments. However, today's complex peace operations should be scrutinized as learning environments in military pedagogy. Further research on the theme is needed in order to develop crisis management operations and various educational and training courses related to them.

As Nørgaard and Holsting (2006) have argued, both combat and contact skills are important in today's peace operations. Wherever the combat skills are needed to maintain security, the contact skills are required to keep contacts with local people and build trust in the peace following the conflict. Learning contact skills takes more time and experience than learning combat skills, which cover a great deal of professional military education at various levels.

Human security concept

New tools in crisis management have been motivated under the umbrella of a relatively new formulation of security called "human security". The human security perspective emphasizes the need to guarantee the safety of the civilian population. This means that security questions should not only be analyzed at the national or international level – one must take into account that the micro-perspective may reveal a lack of security at the individual level.

Human security should be a universal concern. Its components are interdependent, and early prevention is an important part of guaranteeing human security, which as a concept is people-centered (United Nations Development Program, 1994). Human security means safety from chronic threats as well as protection from sudden and hurtful disruptions in daily life (ibid.). The human security concept was launched to a wider audience in the UN Development Report in 1994. It has become one of the key ideas on which the UN and EU have been developing their security policies. According to Nuruzzaman (2006), the human security paradigm in security studies and international relations in the early 1990s signified a paradigm shift in security thinking. The most important factor strengthening its popularity was the changed role of threats at the end of Cold War (ibid.).

The idea of human security altered the model of nation-state-based security. However, it has remained a controversial idea. There are two basic interpretations of the concept. According to a broad, inclusive view, human security requires attenuation of a wide range of threats to the wellbeing of individuals and communities. A narrow view on human security would limit it to the attenuation of the threat of physical violence (Henk, 2007).

According to MacFarlane and Khong (2006), although there are the "development first" and the "security first" camps in human security discussion, these two positions are not mutually exclusive. The co-existence of both dimensions, i.e., development and security, may be seen as important factors in peace processes and development. Because today's conflicts often take place among the civilians, their protection and position in warfare must be analyzed in a coherent way, and this kind of research should have an impact on the education and training of the officers as well.

The human security paradigm lacks a consensual definition, and it was characterized by a large number of institutional and academic views (Nuruzzaman, 2006). Although there are several definitions of human security, the challenges of human security related questions should not be forgotten. As Ewan (2007) emphasizes, more attention should be involved to "bottom-up" and "people-centered" views. The ways to "produce" human security must be studied in a systematic way, and the views of the individuals living under threat in conflict or post-conflict areas need to be acknowledged. From their perspectives, the micro-level and local success of crisis management operations are important factors, too. Therefore, it is interesting to get information on the future officers' views on human security. In military crisis management, the role of professional officers is important and their attitudes may have a strong impact on how various processes within peace operations are carried out.

Human security principle has been a basis for completing the EU's Security Strategy. A working group proposed that a "Human Security Response Force" would consist of 15'000 persons of whom at least one third would be civilians (Study Group on Europe's Security Capabilities, 2004). Beside the goals for military crisis management of the EU, the goals and realization of the civilian crisis management are assessed regularly as well. The security policy of the EU has started to take into account human security and human rights' issues.

Action competence, ethical questions and human security

In military pedagogy, action competence has been developed to integrate psychological, physical, social, cultural, and ethical approaches under the umbrella of human action (Nissinen, 2007). According to Toiskallio (2004), who has developed the framework of action competence, this approach underscores the meaningful role of individuals in military organizations without being individualistic.

Action competence is used most of all at the individual level. There are no efficient units without action competent individuals. The ethical dimension of action competence combines the other three dimensions (Toiskallio, 2000b). Some of the basic elements of action competence are situational skills. Between these situational skills and action competence, there is a mid-level called practical competence (Toiskallio, 2000c).

The central question in military ethics is: Can the killing of a human being ever be ethically acceptable (Stadler, 2003)? Questions related to an individual soldier's decisions as well as to the role of the military in a society are topics in military ethics. War and the final questions deliver relevant conceptions of ethics. The question of military order is a crucial one in a liberal society (ibid). Stadler (ibid.) considers the civilianization of the military an important task. Regarding human security and protection of civilians in armed conflict, soldiers may need to make decisions regarding the use of force in order to protect civilians. Therefore, the human security concept may bring ethical dilemmas which are not easily answered although they are well-known from the history of warfare.

The authority should be combined with responsibility in an adequate way, and this kind of social organization can consist of autonomous individuals (ibid.). In warlike situations, circumstances are problematic, and carrying double chains of responsibility seems to be the most effective way to organize the military (ibid.).

In the military sphere, the factor of responsibility reaches the most serious levels when the leaders have to put their own lives as well as their subordinates' lives as risk. In war or conflict the humanitarian principles of international law have to be observed (Freistetter, 2003). The dilemma may be the controversy between obedience and conscience, and can lead soldiers both to the best as well as to the worst decisions in moral terms (ibid.).

At least theoretically, civilian casualties in warfare can be kept to the absolute minimum due to the advancements in weapon technology. This should at least be the case in warfare which is waged by the high-tech countries (Wirtz, 2003). Those that have the technological advantage utilize it in order to minimize the human losses in their armed forces, too, as it is the case in the military crisis management. However, even from a cognitive perspective, warfare of insurgents, terrorists or guerrillas among local inhabitants may be difficult to handle for an individual soldier in crisis management, because the armed enemy takes action in a hidden manner among civilians.

The ethical challenges in crisis management not only extend to technological questions, because the civil-military relations have become a substantial issue in crisis management. In peacekeeping and crisis management one of the main objectives is to protect the civilians and to enable the peaceful development of the region. Traditional peace-keeping can be described as one type of implementation of human security, although this term did not exist at the time of the first UN led operations.

Mikkonen (2008) has studied the ethical dimension of action competence. Military personnel having work experience in military crisis management had diverse experiences of crisis management work requiring ethical thinking. Decision-making requiring ethical thinking were divided into following groups (Mikkonen, 2008):

- endangering one's own troops or oneself in order to accomplish a particular task or to aid civilians,
- decisions to save local civilians' lives or to distribute humanitarian aid,
- decisions to use armed force,
- exceeding one's powers to reach a goal, and
- leader's remarkable military, administrative, or disciplinary decisions regarding his own unit.

According to Lantto (2005), Finnish conscripts mostly have to develop their views and interpretations on regulations and laws on warfare by their own. In the training of the conscripts, ethical questions are only marginally raised (ibid.). The instructors do not raise these kinds of questions although the conscripts believe that their thinking could be influenced by training (ibid.). This finding raises the question whether the instructors' education and training should be more focused on the didactic dimension regarding the justified existence of national defense and the complicated phenomena related to military crisis management and warfare. The ethical assessment is part of thinking about these kinds of phenomena.

Toiskallio (2005) questions whether we are able to train and educate expert soldiers with a strong sense of personal independence and with a high level of ethical awareness. Especially new challenges, including international crisis management missions based on human security principles, make us think how soldiers' and officers' education and training should be developed in order to respond to these challenges.

Complexity and dynamism are serious requirements for education in the military (Hartmann, 2000). The educational task of superiors is to organize the environment of self-organized and self-reflective learning activities and to be moderators or specialists providing needed information in the process (Hartmann, 2000). From the pedagogic perspective, according to Hartmann, it is important to improve:

- the political education in the military
- the courage of the soldiers to act in spite of complexity
- the ability to learn to qualify and educate him- or herself in military situations
- the adaptability of the military organizations

There is an ethical dimension in human security. In a certain way the human security approach enlarges the views traditionally defined as part of security policy. The human security approach involves taking every civilian's rights into account and it is connected to the universal human rights. However, the international humanitarian law, which is based on the idea of protecting civilians and prisoners of war, has been part of the soldiers' education even before the human security concept evolved. At the Finnish National Defence University human security is taught not as a separate subject. But the theme of human security is handled during several courses – at least in courses on international law and crisis management.

A soldier working as a peacekeeper has to take into account the impacts of his actions both in immediate and longer lasting terms. A soldier should be able to foresee a certain amount of results – even of possibly unintentional character – of his or her actions. The human security approach is relevant in crisis management but it can be applied to the national defense questions, too.

Research question and the questionnaire on the cadets' human security thinking at the Finnish National Defence University

The research question of the study was following: What do the cadets think about human security and what have they learnt about the issue during their studies?

In December 2007, the empirical part of the study was carried out. The fourth-year cadets of the Finnish National Defence University were asked to answer three questions related to human security and human security education. A minority of respondents attending the fourth year cadets' education were completing their studies to fulfill the criteria of a university level examination. There were altogether 123 fourth year students at the National Defence University at that time. At the beginning of the questionnaire, there was a short definition of human security, since the aim of the study was not to assess whether the concept was known by the respondents or not. The cadets answered to the following questions:

- How do you assess the human security concept as it has emerged during your studies at the National Defence University?
- What kind of value for application and what kind of meaning does human security have for your future career as an officer?
- What kind of meaning should human security have in the development of the international crisis management?

No fixed answers were given as alternative choices to these questions in the questionnaire. Therefore, the questions were open to several interpretations, and the answers of the respondents were spontaneous. At the beginning of the questionnaire, there was a brief introduction to the concept of human security.

Results

Altogether 109 respondents completed the questionnaire. Most of them were fourth year cadets who had started their studies at the National Defence University soon after they had finished their military conscript service. But eleven of the respondents were warrant officers who attended the fourth year cadets' education in order to get a university level officer examination. Seven of the respondents were female and none of them were warrant officers. The respondents were born between 1967 and 1984, and the cadets without warrant officer background were born in 1978 or later.

The cadets' general views on human security were analyzed in the answers to the question, "How do you assess the human security concept having been emerged during your studies at the National Defence University?". The concept of human security was not well known by the respondents. 52% of them (57 respondents) expressed that they had not heard about this concept before. 30% of the respondents (33 respondents) expressed clearly that they had learnt something on human security during their studies. 14% (15 respondents) had learnt at least something on the topic. 3% (3 respondents) expressed some thoughts related to the everyday life of the officers. However, the questions relevant to the human security implementation seemed to be acknowledged by most of the respondents, although only a minority of them had been aware of the term human security.

There are no human security classes in the curriculum of the cadets. In various courses, the teachers may have paid attention to the human security. The theme may have been covered without mentioning the concept of human security. Therefore, the process of learning about human security is dependent on the practical outcome of the single courses the cadets attend.

As responses to the question, "What kind of value for application and what kind of meaning does human security have for your future career as an officer?" the relevance of the concept was estimated in various ways by the respondents. The concept was assessed as relevant by most of them, and only a few of them had difficulties to estimate whether this concept might be substantial during their officer career. Mostly the cadets' assessments regarding the meaning of human security in their future careers were positive. Only five respondents considered that they would neither be able to apply the concept to their work nor foresee situations in which they could utilize the concept in any ways. But one respondent considered that the term was mostly "empty words", and there were a few other negative estimations of the concept.

Of those considering human security as an applicable concept, a great deal emphasized the meaning of human security in international crisis management and peacekeeping. In the field of the air forces, security work in order to hinder accidents was considered as promotion of human security by a few respondents.

As the final question, it was asked what kind of meaning human security should have in the development of the international crisis management. There was almost a consensus about the importance of human security in international crisis management, and mostly the human security concept was assessed as relevant in international crisis management. Of the respondents, 27 emphasized human security in the context of protecting the civilians and 68 thought that the concept was of importance (these categories were not inclusive). Independent of the former alternatives, 8 of the respondents thought that the concept would be crucial from the personnel's perspectives, and apart from the former alternatives, the education on human security was emphasized. Two of the respondents thought that the human security concept would be of no or almost of no importance and two of the respondents did not express any opinion on the question.

The causalities between human and state security were seen in various ways. Most of the respondents did not comment on this topic, which was not especially asked in the questionnaire. But some of the respondents considered that the state security would be an outcome of human security – and some others that state security would cause human security. This spontaneous discrepancy in the respondents' thinking may reveal some difference in security thinking which should be studied in a more precise way in the future.

Conclusions

New kinds of conflicts and wars as well as operations other than war have been changing military education and training. Meanwhile, quality requirements of the armed forces are growing and becoming more complex. Military pedagogy is a good basis to enhance education and learning processes which people serving in armed forces need today – and will need in the future (Toiskallio, 2000a).

The tasks of the National Defence Forces comprise national defense, co-operation with various officials, and international crisis management. For historical reasons, national defense has played the most crucial role in the task of the Finnish National Defence Forces. However, today the military crisis management plays a more important role than it used to do. Many of today's armed conflicts are complicated and the personnel educated and trained to work in military crisis management must be prepared for complicated tasks, possibly raising ethical questions.

New conflicts require modification of the education and training. Today's cadets in Finland attend the courses on crisis management. However, should the ethical dimension be added more strongly and in a mainstreaming way into this education and training because the human security approach has implications on ethics as well? Preparation for both national defense and international crisis management operations will probably take place in the armed forces in the future.

National defense and international crisis management tasks represent slightly different paradigms: national defense is based on the traditional paradigm of national security, while crisis management should rely on the idea of human security, underlining each individual's rights for human dignity. The two paradigms are not necessarily mutually exclusive, and in my opinion they can be combined. This is an ethically remarkable task: the warfare role of the armed forces relies on the national security concept, and the crisis management role relies mostly on human security. Of course, international law must be respected in national defense – and the military crisis management personnel must be able to use force in order to protect peace.

Some didactic innovations may take place and perhaps, resolving dilemmas could comprise a part of ethical education of the personnel training for crisis management missions. Dilemmas could be taken as part of the education as ethical, practical, and innovative problem solving regarding work in conflict areas within various cultural contexts. Because plenty of today's cadets will probably work in crisis management operations, the academic level of this type of education could evolve, too.

Since the 1990s military pedagogy has been evolving in Finland and several other countries. The theoretical framework of military pedagogy (MPED) has been broadening. However, the need for military-pedagogical research is not decreasing. The theoretical work should be continued and even more empirical research connecting MPED to other research fields should be made. Other fields of study could inspire both empirical and theoretical research in military pedagogy.

The framework of action competence will remain crucial in MPED in the future. In armed forces, strong ethical competence is needed: complicated operations in military crisis management, cooperation skills required from each soldier, and high-tech weaponry systems may make soldiers' decision-making even more complex than it used to be.

One may raise the question how much human security questions are only practical ones and how much the concept could change thinking – if it can do it at all. I presume that human security has a larger meaning than just giving a new name to the rights acknowledged in the international humanitarian law. The Finnish cadets mostly expressed positive attitudes towards human security although their human security education was narrow. The results support the idea that the amount of human security education of the cadets could be increased in order to respond to the challenges of the crisis management both today and tomorrow.

New concepts may reinforce new ways of thinking. This may be the case regarding the cadets and human security education. Human security education could strengthen the human rights–based thinking of the cadets and future officers. Systematic efforts could take place in order to develop human security education in various countries. Only after these efforts have taken place, we might see how much this education actually influences the cadets' thinking.

Bibliography

Annen, H. (2000). Military Pedagogy in the Swiss Armed Forces.? In J. Toiskallio (Ed.), *Mapping military pedagogy in Europe* (pp. 33-44). Helsinki: National Defence University.

Ewan, P. (2007). Deepening the human security debate: Beyond the politics of conceptual clarification. *Politics, 27* (3), 182-189.

Freistetter, W. (2003). Conscience and authority – virtues and pitfalls of military obedience. In E. Micewski (Ed.), *Civil-military aspects of military ethics* (pp. 53-66). Vienna: Publication Series of the National Defence Academy.

Hartmann, U. (2000). Military Pedagogy in Germany. In J. Toiskallio (Ed.), *Mapping military pedagogy in Europe* (pp. 23-32). Helsinki: National Defence University.

Henk, D. (2007). Human security and the military in the 21st century. In J Toiskallio (Ed.), *Ethical education in the military: What, how and why in the 21st century?* (pp. 211-237). Helsinki: Edita.

Huhtinen, A.-M. (2002). *Sotilasjohtamisen tutkimuksen tieteenfilosofiset perusteet ja menetelmät.* Helsinki: National Defence University.

Lantto, S. (2005). *Puolustusvoimien upseerikouluttajien kasvattajuus varusmiehen näkökulmasta.* Helsinki: National Defence University.

MacFarlane, S. & Khong, Y. (2006). *Human security and the UN. A critical history.* Bloomington: Indiana University Press.

Mikkonen, R. (2008). *Sotilaan eettinen toimintakyky ja päätöksenteko. Teoreettinen mallinnus ja empiirinen tutkimus kriisinhallintaympäristössä.* Helsinki: National Defence University, Department of Education, Series 2, 20/2008.

Nørgaard,K. & Holsting, V. (2006). *International Operations in FOKUS.* Copenhagen: Royal Danish Defence College.

Nuruzzaman, M. (2006). Paradigms in conflict. The contested claims of human security, critical theory and feminism. *Cooperation and conflict: Journal of the Nordic international studies association. 41* (3), 285-303.

Perry, D. (2005). How ethics is taught at the U.S. Army War College. In E. Micewski & H. Annen (Eds.), *Military ethics in professional military education – revisited* (pp. 152-171). Frankfurt am Main: Peter Lang.

Schunk, L. & Nielsson, L. (2002). Danish approach to military pedagogy. In H. Florian (Ed.), *Military pedagogy – an international survey* (pp. 11-28). Frankfurt am Main: Peter Lang.

Stadler, C. (2003).Military ethics as part of a general system of ethics. In E. Micewski (Ed.), *Civil-military aspects of military ethics* (pp. 3-12). Vienna: Publication Series of the National Defence Academy.

Study Group on Europe's Security Capabilities (2004). *A Human Security Doctrine for Europe: the Barcelona Report of the Study Group on Europe's Security Capabilities.* Retrieved from http://eprints.lse.ac.uk/40209/.

Toiskallio, J. (1996). Sotilaspedagogiikan lähtökohtia. In J. Toiskallio (Ed.), *Tietoyhteiskunnan koulutuskulttuuri. Sotilaspedagogisen tutkimusohjelman suuntaviivoja* (pp. 55-67). Helsinki: National Defence University.

Toiskallio, J. (2000a) Introduction: Is there a military pedagogy? In J. Toiskallio (Ed.), *Mapping military pedagogy in Europe* (pp. 6-9). Helsinki: National Defence University.

Toiskallio, J. (2000b). Military pedagogy as a practical human science? In J. Toiskallio (Ed.), *Mapping military pedagogy in Europe* (pp. 45-64). Helsinki: National Defence University.

Toiskallio, J. (2000c). Unohdettu ja uudesti syntynyt. In J. Toiskallio (Ed.), *Näkökulmia sotilaspedagogiseen tutkimukseen* (pp. 33-44). Helsinki: National Defence University.

Toiskallio, J. (2002). Being a soldier in 2020? In J. Toiskallio, W. Royl, R. Heinonen & P. Halonen (Eds.), *Cultures, values and future soldiers* (pp. 97-126). Helsinki: National Defence University.

Toiskallio, J. (2004). Action competence approach to transforming soldiership. In J. Toiskallio (Ed.), *Identity, Ethics and Soldiership* (pp. 107-130). Helsinki: National Defence College.

Toiskallio, J. (2005). Military ethics and action competence. In E. Micewski & D. Pfarr (Eds.), *Civil-military aspects of military ethics. Vol. 2* (pp. 132-143). Vienna: Publication Series of the National Defence Academy.

United Nations Development Program (1994). *Human Development Report.* Retrieved from http://hdr.undp.org/en/reports/global/hdr1994/.

Wirtz, J. (2003). Ethics and the return of the strategy. In E. Micewski (Ed.), *Civil-military aspects of military ethics* (pp. 25-39). Vienna: Publication Series of the National Defence Academy.

3. National Differences and Necessities in Military Pedagogy Acting and Thinking

Gavril Maloş & Adriana Rişnoveanu

The Diagnosis of the Military Learning System through the SWOT Method

Introduction

In this paper we initiate a diagnosis of the military learning system, starting from the present realities of the Romanian military system. This attempt implies using a method of analysis, the most well-known being the SWOT (Strengths, Weaknesses, Opportunities, Threats) method of system analysis, at present the most frequently used technique for system analyses. The first stage in this analysis is that of identifying the strong points of the military learning system in Romania, while the other aspects will be detailed in future papers.

The SWOT technique starts from the assumption that, in our case, a series of internal and external factors act upon the military learning system, and these factors influence its functionality in a positive or negative manner. The four components of the SWOT technique can be defined as follows:

- **Strengths**: the elements that are characteristic to the system that can support its development actions from the inside;
- **Weaknesses**: the factors that destabilize the system from the inside;
- **Opportunities**: the external factors that, by acting upon the system, determine positive effects within it;
- **Threats**: the external factors that generate dysfunctions on the level of processes and relations within the system.

As far as the military learning system is concerned, the SWOT analysis can generate the following elements (Table 1):

Table 1. *SWOT Analysis of the Romanian Military Learning System.*

STRENGTHS	WEAKNESSES
➢ The accreditation of military academic learning institutions according to national standards and procedures	➢ The renown of teachers on an internal and international level is quite low
➢ The scientific research activity that takes place in the military academic learning institutions	➢ A certain inertia of the practices in the system and people's resistance towards change
➢ Institutional cooperation on the level of military academic learning institutions	➢ The lack of a system in which the diplomas offered in the military learning system are recognized in the civil system, and implicitly the

➢ The training abroad of the didactic or auxiliary didactic personnel ➢ The position towards the new educational technologies – e-Learning, Distance Learning ➢ Adopting new national regulations in the field on the level of military institutions ➢ The development of institutions and teaching, learning, and evaluation programs for foreign language knowledge within the army	recognition of the diplomas on the civil labor market
OPPORTUNITIES	**THREATS**
➢ The more pronounced accent placed on the quality of the human resources, the instruction, and the improvement of militaries ➢ New needs for instruction, determined by the new security context and international cooperation ➢ The growing interest for the geopolitical area of the Balkans, which generates partnerships in the field of education and instruction between the main actors on a regional field ➢ Romania has gained a certain know-how with regard to its experience in combat and can become a disseminator of field tested best practices and of lessons learned	➢ The more pronounced attractiveness of the certificates offered by the civil learning system, especially due to their recognition on the civil labor market. ➢ The competition that arises between the military profession and other professions, and the potential decrease in job attractiveness

Diagnosis of the Field of National Defence

This era will be remembered in the modern history of Romania without a doubt as a time of major political, social, and economic change that will leave a fundamental mark on the destiny of our society at the dawn of the third millennium. The evolutionary sense of these changes was represented by joining NATO and EU as well as by Romania gaining back its well deserved place among the democratic and prosperous states of the world.

Romania is developing its own profile in the geostrategic and military context, in the conditions of a real economic development, thus taking part in the effort of collective defense within the North-Atlantic Alliance, in the construction of the military dimension of the EU, as well as in a whole range of missions aimed at furthering the cooperation and partnership engagements assumed on the international level.

Strong points of the defense system

- A well organized and disciplined socio-human environment;
- The confidence that the citizens place in the military organism;
- Permanent perfecting of the legal framework in the field of national defence and of the way this is applied in the process of re-organizing and restructuring the military structures;
- The reorganization of the Ministry of Defence takes place concurrently with the transformation and adaptation of the Alliance to the new challenges imposed by the diverse and fluid security environment;
- The creation and operationalization of structures that can be dislocated, with the aim of participating in the whole range of NATO activities;
- The contribution to regional security and active involvement in the processes of the European Security Policy;
- The development of the special bilateral and multilateral partnerships with the EU defense structures;
- The development of the initiatives for regional and sub-regional military cooperation, as a complementary element that is a priority for the EU integration;
- The adoption of methods and mechanisms that are specific to the NATO and EU, by implementing the system of planning, programming, budgeting, evaluating, and improving the management of the financial resources that are allocated;
- Fulfilling the responsibilities assumed by the Romanian state within the common European capabilities;
- Creating and consolidating the body of civil servants and consolidating the body of civil servants who embrace this occupation as a career professionally and politically neutral;
- The selection, preparation, and instruction of military employees and civil servants who work directly with citizens and economic agents, and the expansion of the usage of informatics equipment and technologies;
- Strengthening the control of the civil society on the military organism, ensuring the transparency of the normative acts issued by the Ministry, as well as their harmonization with the EU legislation;
- The simplification of administrative procedures, the elimination of parallels and redundancies in the activity of different structures, the rationalization of administrative procedures and of the circuit of documents;
- The modernization of the internal control and audit system;

Weak points of the defense system

- ➢ Internal communication difficulties that generate a certain resistance to change;
- ➢ Not fulfilling the attributions of one's position according to the competencies and responsibility transferred to the superior positions;
- ➢ Not reaching the optimal level of linguistic training and knowledge of the allied personnel operating procedures;
- ➢ Insufficient and heterogeneous endowment with informatics technology, as well as the neutralizing in the administration of this and of the communications that take place in maximal parameters;
- ➢ The reduced possibilities for financial motivation of the existent personnel and for attracting the young personnel in the body of civil servants.

Transformation of the military system of learning and training for the army personnel

This concept, developed in the transformation strategy of our army, refers to:

- the elaboration of "The Conception of the Transformation of Military Learning" in accordance with the real needs of the Ministry of Defence, similar to the learning system in the states that are NATO and EU members;
- the correlation of the training system through courses in the academic institutions with the needs for training/instruction of the commandments/troops;
- the adaptation of the professional formation and training programs in order to reflect NATO and EU strategy, doctrines, procedures, and standards;
- the development of the linguistic competency of the military and civilian personnel, with priority for the personnel in the units/structures that can be dislocated, from the perspective of the necessity of reaching an instruction and communication level of the Romanian Army personnel that is similar to that of the allies;
- the review of the regulations regarding the management of military learning, as well as those that refer to the selection, training, and professional development of the military personnel.

As the threats and weak points both cause functioning problems for the military system of learning, their identification will help us to elaborate strategies that would lead to the elimination of these negative factors, thus exploiting the opportunities and strong points in the advantage of the system.

We will consider the system of military learning as being strictly represented by the military learning institutions and their activity, as well as by the connections and formal or informal relations that are established among these.

Strengths of the military learning system

These are represented by those elements inside the military system that represent positive influence factors which – used together with the opportunities offered by the environment that is external to the system – can determine effects of diminishing or eliminating weak points.

Among the strong points of the Romanian military learning system there are:

Accreditation of military academic institutions

The accreditation of military academic institutions according to the national standards and procedures represents a success and a proof that these institutions are recognized and integrated in the national network of academic institutions, together with schools that have a tradition in implementing this process on this educational level.

Scientific research activity

The "CAROL I" National Defence University carries out its research activity within three different structures: the Centre for Strategic Defence and Security Studies, the Centre for Tactics Research and the departments within the Faculty for Command and General Staff.

The products resulting from the scientific research projects are made public either through the magazines edited by the university or on conferences, symposiums, or different communication sessions organized by other academic institutions. The university teachers motivate the students to become more involved in scientific research activity, aiming only at students who attend B.A., M.A., PhD or even post-academic studies.

The students' increasing involvement in scientific research activity can take on a whole range of shapes, from papers that promote each stage of the course, to their involvement in larger projects that the university is involved in either as participant or as coordinator. This represents an important step in building a "nursery" of future military leaders who can adapt much more easily to the rapid changes that occur in the national or international military environment.

The development of the capacity for strategic thinking and thus predicting of the capacity for analyzing and synthesizing a large quantity of information, of the will for self-improvement, of the confidence in one's own intellectual capacities by becoming aware of the relational-value related side of the individual personality – these are just a few of the attributes that scientific research can enhance and that are needed by each component of the military system for the purpose of fast and aware adjustment to the environmental changes.

International cooperation on the level of military academic institutions

a) *The participation and organization of international activities in the military educational field are some of the instruments used for achieving international cooperation in the field.*

Each of the academic military institutions have proven, in the last few years, to be quite preoccupied with organizing communication sessions with international participation of the students, the teachers, and other military education professionals of different institutions. The positive effects for people who organize this kind of events are manifested in a twofold sense:

- On one hand, the institution starts to gain international acknowledgement on a regional or global level through the voice of foreign participants.
- On the other hand, the interpersonal relations established on such occasions will open communication and consultation channels for the most varied fields.

The international trends in the educational field, and the way in which these trends are perceived and implemented in the development strategies of the different participating institutions, are more correctly perceived and can generate interesting ideas and debates on subjects related to inter-university collaboration.

b) *The development of common instruction programs and student and teacher exchanges*

This represents two additional ways of approaching the relations of international cooperation in the field of military education. The Technical Military Academy excels in this field with the organization of PhD courses in partnership with technical universities in France, the Netherlands, and Germany.

The application schools of the categories of army forces have also developed instruction programs for the common instruction of Romanian military and their colleagues in other armies.

c) *The involvement of military academic institutions in the training of the personnel of cooperating partner armies*

This represents a very important means for promoting the reputation of the institution, the army, and of Romania abroad. The majority of the military academic institutions have shown their openness towards such an activity by offering a very wide range of courses to the partners, this offer being present in a courses catalogue published by the Direction for Management and Human Resources at the beginning of each calendar year. This catalogue can also be found on its website. From the perspective of the interest shown by the foreign military institutions towards the training of their own personnel in the military institutions in Romania, these were oriented towards the study programs in English organized by the "CAROL I" National Defence University, the Application School for Naval Forces, the Application School for Combat Units, and the Technical Military Academy.

Training the didactic or auxiliary didactic personnel abroad

Each of us may have had, or wishes to have had a certain type of training in an international setting, and we believe that we have experienced its benefits from both perspectives. Besides the purely technical gains that are related to our daily professional interests, most of the times there are also those of social nature, which are of an incommensurable value. Either they refer to the relationship that develops between the participants or they refer to the extra-curricular social relationship that develop on the level of the host country's society and the values that characterize it. We thus have the possibility of making a comparison and questioning ourselves about what really defines our own sense of moral and ethical values. The facilitation of the adaptation to the working environment in a multicultural environment and the development of tolerance through the capacity of accepting the principle of the diversity of opinions represent only two elements that may increase the professional performance of the military teachers and can be developed by participating in training programs abroad.

We consider that the didactic and auxiliary didactic personnel in military academic institutions have to occupy a priority position when it comes to the participation of Ministry of Defence personnel in courses abroad. It is good that this is mentioned also in the documents that regulate the activity of international cooperation in the field of education and military instruction.

Position towards new learning technologies – e-Learning, Distance Learning

In a general sense, e-learning refers to the sum of educational situations that use in a significant manner the means of communication and information technology. Seen from the perspective offered by this definition, many of the didactic actions that take place in the military academic institutions can be placed in the area of e-learning. We can find more and more often electronic didactic materials such as presentations in different formats, maps, didactical movies, dictionaries, encyclopedias, etc. in military academic institutions.

Distance Learning represents a planned teaching-learning experience organized by an institution that mediates the provision of materials in a sequential and logical order so that they can be assimilated by students in a personal manner. The difference between the two concepts – e-learning and distance learning – consists of the means of mediation they use. In a particular sense, e-learning represents a type of distance education in which mediation is realized through the new communication and information technologies – especially the internet.

As far as the activities related to using this new form of supplying education are concerned, we will give a few examples that reflect the accomplishments of the military institutions in this field, such as:

- The establishment of a Department for Open Distance Learning in the structure of the National Defence University;
- The agreement between the "CAROL I" National Defence University and the NATO College for Defence in Rome on a collaboration protocol that would give the Romanian participants access to the online courses of the NATO learning institution;
- The organization of a set of online courses at the "CAROL I" National Defence University (starting with the 2005-2006 academic year), such as: the NATO Introduction Course, the Course regarding the European Security Policy, the Crisis Management Course, etc.
- The e-learning program of the Academy of Terrestrial Forces includes the project of informatics system for the management of learning and projects for online courses. So far eight projects for such courses were made and four more were adopted from the PfP Consortium of Defence Academies.

Punctual actions such as those mentioned above have to be extended on the level of the whole military learning system by integrating them in a new educational policy that ensures the militaries' access to education, from wherever they are and whenever they consider they are (psychologically and physically) ready for an educational activity.

Adoption of new national regulations on the level of the academic military institutions

The introduction of the 3 cycles of university studies, i.e., B.A., M.A. and PhD, according to Law no. 288/2004 regarding the organization of university studies, refers to the alignment of Romanian academic learning to the trends and development of European learning. Its aim, besides the facilitation of mobility for students, was to ensure unity on European level as far as the "portrait" of the work force in Europe is concerned. This was entirely accomplished at the "CAROL I" National Defence University and the Technical Military Academy, and only partially at the "Mircea cel Bătrân" Naval Academy.

As far as *ensuring the quality of education* is concerned, shortly after the regulations in this field appeared, the senate of the "CAROL I" National Defence University adopted the following documents: the strategic plan of the "Carol I" National Defence University for the period 01.10.2005-30.09.2010, the operational plan for the 2005-2006 academic year and the system for ensuring the quality of the educational services in the "CAROL I" National Defence University. The fact that these plans were elaborated and approved quite fast indicates the confidence the academic institution has in the concept of quality assurance on one hand, and the managerial dynamism on the level of this institution on the other.

The democratic managerial style of people who are part of the management of the military academic institutions, exemplified by the confidence that each subordinate has something good to offer to the institution, has created a favorable climate for developing a desire for self-improvement on the level of the whole institution.

The establishment of the structures (e.g., the university ethics commissions) and the elaboration of documents (e.g., the university ethics code) for the promotion and monitoring of the principles of the *university ethics* on the level of military academic institutions offers us an optimistic perspective on the quality of the relationships established on the level of academic communities. The Academy of Terrestrial Forces and the Naval Academy are two of the military academic institutions that have established their principles, norms, and regulations that define the university ethics.

The promotion of the military academic institutions' image on the internal and international level is still quite reserved, first of all because of the inexistence of a unitary conception in this field, and secondly due to the very high costs that are involved in the design, purchase, and dissemination of promotional products and actions.

Development of institutions and programs of teaching, learning, and evaluating the foreign language knowledge in the army

If we are to refer strictly to the involvement of academic institutions, we need to point out, first of all, the exceptional quality of the teachers and their degree of involvement in defining and applying the measures adopted on the level of decision-making structures. The relations that were established with international experts specialized in the elaboration of didactic material and evaluating instruments, the involvement of the teachers in increasing the awareness of the role that the knowledge of foreign languages plays in the professional development of each military – these are two elements that make us confident towards learning foreign languages.

The teachers' training abroad was realized in time through their participation of different courses, seminaries, and conferences organized in military or civil institutions in the USA, Great Britain, Canada, France, Germany, Austria, Hungary, Turkey, Estonia, Slovenia, Bulgaria, etc. The number of participants in the training activities abroad for foreign languages teachers has increased significantly in the last four years. While there were 14 graduates of studies abroad among the foreign languages teachers in 2000, their number was already doubled in 2003 and almost tripled in 2006, reaching 39.

As for the training in Romania, this takes place through methods courses organized by the "CAROL I" National Defence University and the "Mircea cel Bătrân" Naval Academy. So far two such programs were organized – one had the topic "Getting familiar with the STANAG 6001 requests and their ways of application in the military foreign languages learning", with 130 teachers participating in the 2004-2007 period, and the second with the topic "The method of teaching the NATO operational terminology", with 66 participants in the 2005-2007 period.

Conclusions

This paper approaches, in a first analysis, the strong points of the military learning system. The other aspects of the SWOT analysis will be discussed in other papers. The SWOT analysis can synthesize the key points of a didactical approach because:

- categorizing the problems and advantages according to the four SWOT categories allows a much easier identification of a strategy and of ways of developing a modern CV, which can be adapted to reform its demands more quickly;
- the method can be easily adapted to the specific needs of different category activities in the Romanian learning system, in a bottom-up approach;

- the SWOT analysis is a concrete way of identifying the *strengths* and *weaknesses*, and of defining the *opportunities* and *threats* the military learning establishment is confronted with, both in the didactic approach as well as in the methodology for shaping teachers.
- it allows a constructive and efficient evaluation/self-evaluation of the strategy applied on a macro level.

Bibliography

Annen, H. & Royl, W. (Eds.). (2007). *Military Pedagogy in Progress – Studies for Military Pedagogy, Military Science & Security Policy, Vol. 10*. Frankfurt am Main: Peter Lang.

Buckingham, M. & Goffman, C. (2005). *Manager contra curentului*. Bukarest: Editura ALLFA.

Ciobanu, O., Pescaru A. & Păduraru M. (2004). *Pedagogie*. Bukarest: Editura ASE.

Cole, G. (2000). *Managementul personalului*. Bukarest: Editura Codecs.

Diaconu, M. (2004). *Sociologia educaţiei*. Bukarest: Editura ASE.

Drucker, P. (2005). *Despre profesia de manager*. Bukarest: Editura Meteor Press.

Druţă, M. (2001). *Iniţiere în psihologia educaţiei*. Bukarest: Editura ASE.

Goleman, D. (2001). *Inteligenţa emoţională*. Bukarest: Editura Curtea Veche.

Istrate, E., Rîşnoveanu, A. & Vasiliu, V. (2005). *Didactica specialităţii*. Bukarest: Editura Universităţii Naţionale de Apărare "Carol I".

Jinga, I. (2001). *Managementul învăţământului*. Bukarest: Editura Aldin.

Maloş, G. (2005). *System Engineering in Acquisition Management for Defence*. Bukarest: "Carol I" National Defence University Publishing House.

Mardar, S. & Maloş, G. (Eds.). (2007). *Educating and Training Officers for Interoperability. The 7th International Conference on Military Pedagogy*. Bucharest: "Carol I" National Defence University Publishing House.

Peter Foot

Educating for Security: The Challenge to Military Academies[1]

It *ought* to be clear by now. Professional military education could reasonably be expected, among its practitioners at least, to be the subject of a widespread consensus. The difficult but nonetheless probable recovery of America's international moral authority after George W. Bush[2]; the rise of Russia as a skillfully ruthless player often at odds with its own interests; the economic-strategic mismatch of rising powers such as China and India; the continued geopolitical fading of Europe, even as efforts are made by the EU to bolster its significance on the world stage – all these are evidence of the new setting. More so, there is a consensus that no military activity on its own will solve the kind of issues the international community now is confronted with in failed and failing states. There is agreement that combat units are necessary, but not nearly sufficient, and that the real job, in the long run, has to be done by civilian agencies of one kind or another. From this one would surely expect that few would disagree with the notion that military training and education needs to take the wider political, economic, social, ethnic and religious dimensions fully into account before planning the deployment of military units.

Yet – as is so often the case – institutional inertia is very powerful in forestalling change or inhibiting reform. Also, the delay in reform is reinforced by the military profession's naturally respectful view of itself. Political scientist Samuel P. Huntington spoke for generations of modern officers when he characterized the ideal officer as patriotic, yet almost above patriotism, in the sense of being a part of the brotherhood of arms, a member of a profession that shares characteristics with the other great professions, but remains distinct from them (1964). He argued that there is an unchanging quality about leadership and military expertise:

> The peculiar skill of the military officer is universal in the sense that its essence is not affected by changes in time or location. Just as the qualifications of a good surgeon are the same in Zurich as they are in New York, the same standards of professional military competence apply in Russia as in America and in the nineteenth century as in the twentieth. (p. 13)

---

1   This essay is a written-up version of a verbal presentation given at the 2008 IAMP conference in Helsinki.
2   No foreseeable US Secretary of State will repeat Madeleine Albright's words of 19 Feb 1998: "If we have to use force, it is because we are America! We are the indispensable nation. We stand tall. We see further into the future."

That 'peculiar skill' – the professional art – is a deeply valued possession to those who hold it. Together with their acceptance of personally unlimited liability, they see it not unreasonably as something setting them apart. Providing the next generation with the keys to higher professionalism by way of staff and war college graduation becomes a constant and inviolable, even sacred, responsibility. This line of reasoning will insist that no one outside that group can either know what is required or have the skills to infuse the men and women in uniform selected for higher command in the service of their country with the necessary values.

It is not in any way to belittle such conservative views by pointing out that this is not the whole story. There is a wider perspective than the one that is internal to the profession of arms and its principled obligation to pass on the torch of high endeavor. Or as that staunch pro-military commentator Edward Luttwak put it (1995, p. 115), there might be "a profound contradiction between the prevailing military mentality [...] and current contingencies".[3] It is these contradictions that have been well developed by Mary Kaldor in her thesis on the 'new wars' (2006). Current contingencies are often not expressions of a new status quo in the sense of a post-bellum settlement in favor of one side of the conflict. More often, *keeping the conflict going* is the source of the belligerents legitimacy, power and profit. Africa has too many examples of this to mention here. Paramilitaries and armed thugs very often comprise the front line of confronting conventional military forces – a set of conditions that is far removed from the myths of heroic warfare that animate almost all military establishments, especially the ones that inculcate professional values in the impressionable new recruit, officer or otherwise.

It is therefore no wonder that the gaps between doctrine, training, experience and new operations have grown larger. Mistakes do not have to be of the order of those committed by the Israelis in the 2006 Lebanon War, if on a narrower military spectrum than most contemporary crises, but there is a lesson to be learned from that experience for all.[4] Senior officers planned and conducted the war as if it was the classic, ground-based, force-on-force encounter Israelis have been familiar with. Junior officers, on the other hand, had only experienced stone-throwing young men in Ramallah and were disconnected from their superiors as to military experience and purpose. In contrast to both groups, the Israeli chief of staff, General Dan Halutz, was seemingly motivated more by the exaggerated doctrines of air power decisiveness from Afghanistan 2001 and Iraq 2003 than anything else. Needless to say, none of the three mindsets can be said to have understood the nature of the Hezbollah challenge. Despite unwarranted claims about an Israeli defeat, this was more than an embarrassment. It demonstrated the extent of what can go

---

3   Luttwak drew conclusions from this contradiction that are largely relevant to US policy makers.
4   I am grateful to Andrew Sharpe for these insights.

wrong in the post-Cold War world if training, military culture, equipment, doctrine and adequate intelligence are not brought into alignment.

The Doctrine Problem

To a degree, it is with this set of conditions that militaries have been wrestling with for the past 20 years, and especially since the opening of conflicts in the Balkans following the break-up of former Yugoslavia. The possible approaches to the issues at hand were elevated to (what turned out to be temporary) doctrinal golden calves, rather than developed into more enduring strategic thinking. Each of these 'graven images' has made its solemn, if brief, progress through military training establishments. For example, much was made of the Revolution in Military Affairs (RMA) as a technological solution to warfare and problem solving in the last decade of the 20th century. Once Donald Rumsfeld had ordered that peace support operations were to be regarded and funded as equally important as combat operations, such ideas had declining currency. Boyd's OODA loop, and network-centric warfare attracted institutional and some professional loyalties. But by the end of the first decade of the 21st century, these approaches had a certain dated appeal to all but the real enthusiast. The 'Pentagon's New Map' had a similar fate. Effects-Based Operations (EBO) largely fell victim to the internecine warfare about doctrine and funding between the US Armed Forces' service branches; the idea was quietly emasculated by NATO's acceptance of the Effects-Based *Approach* to Operations (EBAO) – which watered down the doctrinal point about achieving 'effects' as the predominant basis for planning and acquisition. The UK's Blair government was most vociferous in advocating a 'joined-up government' approach to operations whereby separations between departments would be swept away by a common understanding of the problem to be tackled collectively and synchronously. The continued commitment to national resilience planning is a significant part of this but, overall, the UK government has the same problems in breaking across departmental 'silos' as any other. Norway has been moving slowly to educate the international community about sending 'integrated missions' into the field – whereby the composition of arriving deputations to a region of crisis is appropriately composed and professionally homogenized. But even within UN agencies there are arguments about roles and missions.

The most recent idea that has run into trouble is that of the Comprehensive Approach, a whimsy of members of both the European Union and NATO (UK Joint Delegation to NATO, 2010):

> Experience from NATO operations has demonstrated to Allies that co-ordination between a wide spectrum of actors from the international community, both military and civilian, is essen-

tial to achieving key objectives of lasting stability and security. The Declaration on Alliance Security, issued in April 2009, underlines the importance of strengthening our cooperation with other international actors, including the UN, the EU, the Organisation for Security and Cooperation in Europe, and the African Union, in order to improve our ability to meet new challenges.

This is positive progress, but it is limited by at least three factors. First, the central position of another uniformed and disciplined agency is nowhere recognized as having a role other than being lumped together with non-military actors. Police are crucial to the re-establishment of the rule of law and have often very difficult problems associated with being handed issues by military authorities: burden sharing and divisions of labor are far from clear and usually vary from situation to situation. Secondly, the police are very different from the plethora of civilian agencies that also have to be taken into consideration (e.g., planning and funding). Finally, while humanitarian and other aid agencies share a hostile consensus about the supposed iniquities of the military in humanitarian or peace support missions, they are also antagonistic to each other when it comes to basking in the limelight, funding and the aid agencies' equivalent of 'market share'.

What Price Experience?

All of this makes the military's professional deliberation about training and education even more difficult than it has ever been since the end of bipolarity. Unsurprisingly, a few senior officers have opted for a crude 'return to basics' approach for the future of their militaries. They have sought to define the military role in a narrow way, which celebrates a Spartan rather than an Athenian approach (Lovell, 1979). A former head of the United States Air Force, General John P. Jumper, perfectly exemplified this approach in 2005, i.e., after post-2003 Iraq invasion had become to be deeply problematic. He ordered that with regard to promotion purposes, postgraduate qualifications were to be ignored (unless one was a medic or chaplain). He himself had at one time acquired an MBA from Golden Gate University and regarded it as having been inconsequential for his career as a fighter pilot. The same, must broadly hold true for other Air Force officers, he felt – or if they did benefit from studies, it would appear as improvements in performance, thus to be rewarded in the usual way. There would be no 'double counting', i.e., no combined promotional benefits for the postgraduate degree *and* the improved performance. Superficially, this may seem fair. But it represented a significant expression of anti-intellectualism at the very moment when disciplined, creative thinking – and not just about air power – was in particular demand. No doubt the authorities at the Air University at Maxwell Air-force Base made that point forcefully but were over-ruled. Sensibly, a paradox has resolved the issue for many. Although the US and other armed forces now have more combat-experienced cadres than at any time since the Second World War, the best military leadership examples are invariably keen to be exposed to the conceptual, historical and contextual settings of combat and related missions. General David Petraeus' employment of four PhDs as part of his counterinsurgency team in Iraq is not an accident or an aberration. The best field officers want education; those more narrowly responsible for military training often get in the way.

One practical response to this conundrum might be the development of an intellectual elite among the officer cadre. Let the best take care of business. Instinctively, viscerally even, this runs too much against the grain of many militaries and thus this path is unlikely to be taken. Another argument is that an elite, no matter how gifted and trained, relies on the staff work of less able colleagues.[5] A disconnect in training and education in that sense is illogical and likely to be counterproductive. A further reason for eschewing this divisive approach is the phenomenon recognized in the phrase 'the strategic corporal' (Krulak, 1999). There is no overwhelming case for depriving NCOs of operational experience that will define such an approach as a success. As the phrase implies, a badly trained corporal can have negative strategic effects because of his ignorance of relevant factors and events. This is overcome by more than just clever drafting of rules of engagement and hoping for the best. The US Marine Corps may have identified the significance of 'the strategic corporal', but one is more hard-pressed to see where training investment in that area has taken place – another example of how institutional culture is not easily changed in practice.

What all of this indicates is the need for a level of education of the entire deployed force, a level that readies it for the complexities and ambiguities that are inevitable in the field. Such an approach also implies that senior officers need to be ready and able to undertake further and challenging education, both for themselves and in order to demonstrate leadership in this area. The strategic climate for this has rarely been better. The fact that none of us can find a decent label for our age other than 'post-Cold War' is a powerful testament of our own honest limitations in perception and analysis. The United States, Germany and Switzerland, to name just three, have invested in senior officer education about strategic and operational realities as currently understood. This is a trend that needs to continue and be followed internationally. Wherever possible, senior officers need to meet those they are going to have to deal with as part of a coalition – the IGOs, NGOs, police, other services and the UN agencies involved. For this to

---

5   'Able' here is defined as having the ability to rise to CHOD rank or command of their service. It does not mean 'ability' in any other, or absolute, sense.

work well, there will be a need for not merely courses, but *exercises* in how to manage things. Cultural exchanges cannot be overvalued in this context – the values of an institution are carried largely in their respective cultures. Bringing diverse cultures together is what is now needed – crisis management, humanitarian aid, infrastructure rebuilding, police, justice, constitutional and military. The effectiveness of military personnel in the field is to a certain extent due to exercises being part of training routines. Police are similarly practiced. Now, civilian agencies and representatives need to embark on the same approach – and in conjunction with the military and police.

Because of the way things are structured in most countries, it makes sense to host such events at military or police colleges. Simply put, they have the facilities needed to turn them into a success. In particular, they have the international contacts necessary to ensure that nothing is done that is too narrowly *national* in character, either in subject matter or methodology. But the overwhelming point is to provide a setting that allows expression of the idea that all parties to such planning and practice cannot manage on their own; all are interdependent. Blame and credit are not divisible. Having a 'good war' was always the goal of journalists, governments and soldiers. This was always measured as an element of some notion of battlefield success or otherwise. Nowadays, a 'good operation' will be measured by how much collaboration took place, how much effectiveness resulted, and how much collective effort contributed to the outcome. It will not always be straightforward, but it is achievable. Essential to all of this is a commitment to critical thinking. We have seen the result of groupthink, the herd instinct of economists and bankers, content to leave developments to fate (i.e, "the invisible hand" of market). The result was the near-catastrophe of global financial breakdown beginning in late 2007. Strategists at least have always understood that they are more than the simple victims of circumstance. They could *do something* to avoid the worst, or make sure things turned out better than would have been the case otherwise. This is the task facing the diverse communities joined in providing for security in the future. This task will not be easy, of course; it would be foolish to claim otherwise. But at least the future can be prepared for. Senior military officers, their decision-making colleagues in the police and other institutions should most definitely take note.

Bibliography

Huntington, S. (1964). *The Soldier and the State: The Theory and Politics of Civil-Military Relations.* Cambridge, MA: Harvard University Press.

Kaldor, M. (2006). *New and Old Wars: Organised Violence in a Global Era.* Cambridge: Polity Press.

Krulak, C. (1999). The Strategic Corporal: Leadership in the Three Block War. *Marines Magazine, January,* 18-22.

Lovell, J. (1979). *Neither Athens Nor Sparta? The American Service Academies in Transition.* Bloomington, IN: University Press.

Luttwak, E. (1995). Towards-Post-Heroic Warfare. *Foreign Affairs, 74*(3), 109-122.

UK Joint Delegation to NATO (2010). *Comprehensive Approach.* Retrieved from http://uknato.fco.gov.uk/en/uk-in-nato/comprehensive-approach

Giuseppe Caforio

The Nature of Security Threats in the Perceptions of Future Civil and Military Elites

Foreword

The profound geopolitical transformations beginning in 1989 led, for a certain number of years, to a decreased perception of threat by European populations (and others as well) that was shortly followed by a sizeable decrease in armed forces and military budgets. But the illusion of the *End of History* (Fukuyama, 1992) and of a *peace dividend* (Mearsheimer, 1990) soon had to give way to an international reality that was much more turbulent than in the past but which, up to the events of September 11, 2001, did not strongly impress public opinion. Threat perception was a datum of so little interest after 1989 that, for instance, the Eurobarometers of the 1990s and up to 2002 did not include this item in their questionnaires. The attacks of September 11, 2001 represented a turning point in a crescendo of preoccupations now registered by opinion polls, which reveal that the countries of the European Community (as well as other countries) have developed sensitivity to individual threats. This trend is depicted graphically in Figure 1.

After 2001, the citizens of the examined European countries thus appear to be concerned by what have been termed the "new security threats", namely international terrorism, proliferation of weapons of mass destruction, organized crime, an accident in a nuclear power station, ethnic conflicts, and a world war. Notwithstanding the limitations of such a short-term analysis (2002-2003), most of these threats prove to be on the rise. Since the fall of the Iron Curtain in 1989, the security problem therefore no longer centers on the necessity of maintaining a balance of forces with the opposing bloc but on the need to establish and/or maintain a peaceful situation in the world. Moreover, since 2001, countering the new menace of Islamic fundamentalism in its most aggressive forms has become an additional challenge. This is also the framework of the exponential growth in the number of peacekeeping missions from the 1980s to 2000s and beyond, missions that have become an important (in some cases major) part of the operational commitment of the armed forces of Europe (Figure 2).

Figure 1. The fears of EU citizens (EU15).

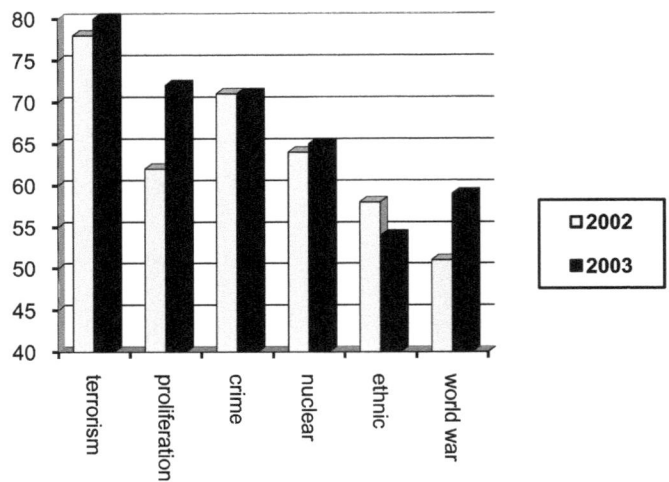

(source: Eurobarometers 57 and 59. Author's elaboration)

Figure 2. UN Peacekeeping Operations Timeline (Source: www.un.org/Depts/dpko/timeline. Author's adaptation).

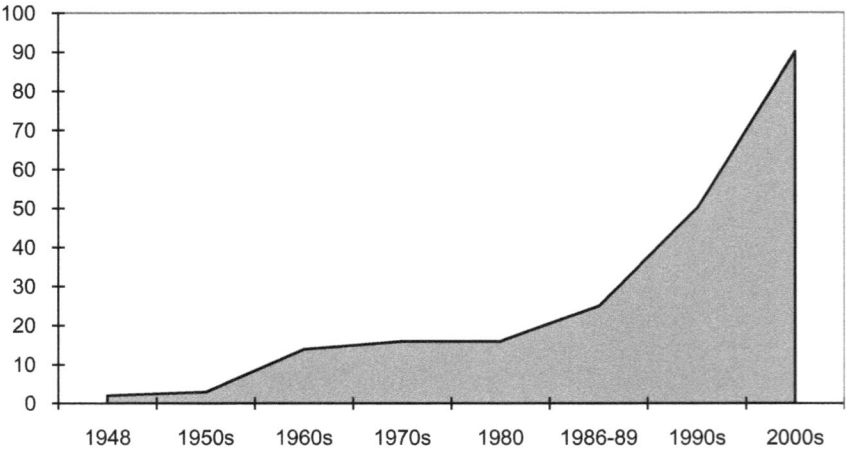

The situation created by the break-up of the Soviet bloc, the resurgence of old and new nationalisms, and the clash of civilizations, ethnic groups and religious beliefs has led to today's society being defined as a risk society, where perceived risk constitutes one of the elements that characterize Western-style societies[1] most (see in this regard Barman, 1999; Beck, 2000, and Giddens, 1991). As Alessia Zanetti for example writes (2004):

> The new threats, arising from deep changes at the international level and generated by a complex of political, economic, technological, social and environmental factors, are no longer ascribable to a single "great enemy" but present themselves in a diffuse, pervasive, horizontal manner, altering the very perception of the future and significantly influencing action strategies. (p. 55)

The social representation of the risks thus depicts societies characterized by a perception of both individual and collective security that is precarious, exposed to a set of threats of international dimensions, allowing scholars to speak of an *international risk society* (Zanetti, 2004). The study and evaluation of the social perception of threats is also important because this perception is not only the result of objective events, but in large measure of collective representations (Maniscalco, 2004) generated by a complex set of factors. In this framework an interesting element of knowledge (or at least of cognitive tendencies) is provided by a survey that a research group coordinated by the author[2] was able to carry out on a significant number of democratic countries that are listed in the next section.

The data of this survey do not pretend to be exhaustive or even complete, but constitute a first approach to knowledge of the perception of threats on the part of the future elites of a fair number of countries at a given historical moment.

Introduction

As announced, this paper is based on data from an empirical survey that was conducted in a good sample of European countries. The sample was comprised of the following countries: Bulgaria, France, Germany, Italy, the Netherlands, Poland, Romania, Slovenia, South Africa, Spain, Sweden, Switzerland, and Turkey. Even if South Africa does not belong to Europe, I have included its data also, since its culture is rather close to European culture. Another interesting case is

---

1   Apart from the Western societies as traditionally defined, also societies that are inspired by such models and are gradually coming closer to them, e.g., the former communist countries of Eastern Europe, Russia, etc.
2   The group consisted of researchers of the Research Committee on "Armed Forces and Conflict Resolution" of the International Sociological Association and members of the ERGOMAS Working Group "The Military Profession".

that of Turkey: as known, this country sits astride the boundary between Europe and Asia, but its current interest in becoming a member of the European Union, as well as the likelihood that, in the medium term, such accession will be achieved, led me to consider it European. It also seems consistent with the secularism and tolerance of European culture to include one with a Muslim majority in addition to countries with Catholic, Protestant and Orthodox majorities. The research was carried out on a broader subject[3], but it still offers significant insights for the subject under review.

The particular significance of the research stems from the fact that it is not a generic opinion poll of a sampling of citizens from each country selected according to the common characteristics of representativeness of ordinary surveys, but a research on samples of students both at civilian universities and at military academies. These "future elites" of the participating countries were surveyed in depth by means of a questionnaire consisting of 45 questions administered during the years 2004/2005. A total of 3003 young people born between 1974 and 1986 were surveyed: the sample composition is shown in Table 1.

Table 1. *Composition of the sample*

	Q20. What are you studying?				
	Military Academies	Various	Economics	Law	Political Science
Number of usable questionnaires	1327	244	599	452	381
Percentages (horizontal)	44.2	8.1	19.9	15.1	12.7
Male %	88.0	44.3	50.7	54.2	49.1
Female %	12.0	55.7	49.3	45.8	50.9
Class year: % 1st and 2nd	58.7	42.6	47.4	51.9	47.4
3rd and higher	41.3	57.4	52.6	48.1	52.6

The survey sheds light on some of the value attitudes of the youths of the examined countries in a given historical moment (thereby also permitting a diachronic comparison in a later analogous investigation) and allows for a comparison of these attitudes among the youths of different geopolitical areas. An aspect that might be considered a limit of this research must be borne in mind, however: it was conducted among those that we have called the "future elites" of the examined countries, i.e., only among young people attending university-level programmes, with the exclusion of those already employed in other occupations. Our point of

---

3   The topic of the research was "Cultural Differences between the Military and Parent Society in Democratic Countries", the results have been published in a book with the same title by Elsevier (2007).

view hinges on the hypothesis that it is the elite that dictates the tendencies their non-academic coevals sooner or later follow.

Research data

In the whole sample examined, the perception of the various threats is as shown in Table 2.

Table 2. Likelihood of threats

Threat likelihood	%
• Armed conflict over the control of vital raw materials	**61.0**
• Mass immigration from foreign countries	**55.6**
• Terrorism	**64.4**
• International drug trafficking	**75.1**
• Organized crime	**79.4**
• Environmental problems (air pollution, etc.)	**63.3**
• A possible indirect involvement in a civil war	30.6
• An armed conflict between African countries with which we have co-operation relations	25.0
• An armed conflict in the Middle East	**50.7**
• Attacks on computer networks	**60.9**
• Nuclear blackmail by Third World countries	31.7
• Nuclear war between Third World countries with global consequences	37.9
• An armed conflict between Asian countries	32.9
• Proliferation of weapons of mass destruction	**62.0**
• An accidental nuclear war	35.2
• Military attack by a foreign state	30.6

In the list of proposed items we can see that a small probability of occurrence is given to a number of threats by the sample as a whole, while others are considered much more likely. For reasons of simplicity and readability the following examination will concentrate on these last items in the resulting order of probability:

1. Organized crime
2. International drug trafficking
3. Terrorism
4. Environmental problems (air pollution, etc.)
5. Proliferation of weapons of mass destruction
6. Armed conflict over the control of vital raw materials
7. Attacks on computer networks
8. Mass immigration from foreign countries
9. An armed conflict in the Middle East

A first reflection that arises from this list of priorities is that those that we might term "police" threats and military threats divide the list in a fairly clear-cut way, since the priorities 1, 2, 4, and 7 are of predominant interest to law enforcement, the priorities 3, 5, 6, and 9 relate to the military, and number 8 concerns both areas in equal measure.

Figure 3. List of priorities

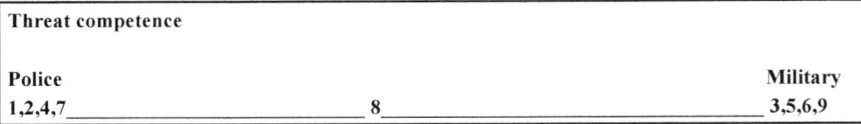

But the level of threat perception changes substantially when we pass from one examined country to the next, as shown in detail in Table 3 below. For each threat the upper number indicates the percentages of probability assigned to it by the future elites of each country. The lower number indicates the order of frequency with which the likelihood of such a threat is seen in each individual country. The last line of the table is an attempt to give a dimension, albeit very approximate, of the global intensity of the perceived threat in each examined country by taking the average of the percentages for the perception of each threat as felt by the young people of that country. Beneath these percentages, a sort of classification by country is proposed, according to the perceived average global threat.

Table 3. Threat percentages by countries (and classification inside each country)

Threats	Bulgaria	France	Germany	Italy	Netherlands	Poland	Romania	Slovenia	South Africa	Spain	Sweden	Switzerland	Turkey
Armed conflict over the control of vital raw materials	64.4 / 6	59.8 / 8	68.1 / 4	36.6 / 9	54.6 / 7	58.2 / 6	83.2 / 5	50.0 / 5	61.6 / 6	62.6 / 2	45.2 / 8	66.9 / 4	78.1 / 5
Mass immigration from foreign countries	47.1 / 8	66.4 / 7	45.1 / 8	84.4 / 3	61.7 / 4	48.2 / 8	39.5 / 9	42.7 / 7	90.0 / 1	52.0 / 7	45.9 / 7	62.6 / 6	47.0 / 8
Terrorism	71.1 / 4	89.1 / 1	58.6 / 5	88.5 / 2	54.0 / 8	44.6 / 9	88.3 / 3	28.0 / 8	68.4 / 5	93.7 / 1	52.4 / 6	26.8 / 9	94.8 / 1
International drug trafficking	87.6 / 1	76.5 / 2	55.7 / 6	78.6 / 4	80.9 / 1	87.9 / 2	94.4 / 1	58.9 / 3	88.8 / 3	55.0 / 5	75.6 / 2	76.1 / 1	81.7 / 3
Organized crime	84.2 / 2	73.5 / 4	71.8 / 2	89.8 / 1	78.7 / 2	94.2 / 1	93.8 / 2	75.5 / 1	89.3 / 2	60.1 / 4	86.4 / 1	67.8 / 3	87.2 / 2
Environmental problems (air pollution, etc.)	70.2 / 5	74.5 / 3	43.8 / 9	55.0 / 7	60.3 / 5	84.6 / 3	84.5 / 4	61.4 / 2	69.5 / 4	45.4 / 8	58.8 / 4	65.9 / 5	65.7 / 6
Armed conflict in the Middle East	42.2 / 9	70.4 / 6	50.7 / 7	62.2 / 5	56.0 / 6	56.8 / 7	47.5 / 8	12.9 / 9	39.5 / 9	53.0 / 6	44.1 / 9	37.8 / 8	78.5 / 4
Attacks on computer networks	53.3 / 7	59.6 / 9	73.6 / 1	45.4 / 8	66.3 / 3	63.5 / 5	73.2 / 7	50.3 / 4	60.1 / 7	36.6 / 9	72.8 / 3	74.7 / 2	46.2 / 9
Proliferation of weapons of mass destruction	72.8 / 3	70.6 / 5	70.7 / 3	57.4 / 6	53.5 / 9	71.1 / 4	79.9 / 6	50.0 / 6	47.8 / 8	61.5 / 3	57.0 / 5	59.4 / 7	60.4 / 7
Average feeling of being threatened	65.9 / 7	71.1 / 2	59.8 / 10	66.1 / 6	62.9 / 8	67.7 / 5	76.0 / 1	47.8 / 13	68.3 / 4	57.8 / 12	59.8 / 9	59.8 / 10	71.0 / 3

Having explained the table, let's now have a look at what the data tell us. Examining the threats that are considered most likely from top to bottom, we see that mass immigration is considered a significant threat by a group of countries that appear most exposed to this phenomenon due to their geographic locations, namely South Africa, Italy, and also the Netherlands. Terrorism is perceived as the major threat in Spain, Turkey and France, but it is also very relevant in Italy and Romania. Drug trafficking and organized crime – two threats that tend to blend into each other – are the ones most chosen by the sample as a whole and are the top priorities for the respondents from Bulgaria, the Netherlands, Poland,

Romania, Slovenia, and Sweden. They are also significant for all the others, although somewhat less so for Spain, Germany, and France. Interestingly, drugs and criminality seem to constitute the biggest problem in all the former communist countries of the sample. The environmental threat is given a fair amount of importance by the Slovenian, Polish and French respondents. The possibility of an attack on computer networks is deemed a priority by the Germans, but is a serious threat also for the Swiss, Swedish, and Dutch. Countries where these networks are most developed and most essential to the associational life of the country seem to feel most vulnerable in that regard. If we want to give credence to the average percentage value of the threat calculated above (last line of the table), we can see that the future elites of Romania, France, Turkey, and South Africa are the ones that are most concerned about their country's security.

Table 4. Perceived threats by status (%).

	Cadets	Students
Armed conflict over the control of vital raw materials	60.0	**61.6**
Mass immigration from foreign countries	**61.6**	50.7
Terrorism	**67.7**	61.8
International drug trafficking	**78.3**	72.4
Organized crime	**81.9**	77.2
Environmental problems (air pollution, etc.)	61.0	**65.0**
An armed conflict in the Middle East	**56.3**	46.1
Attacks on computer networks	**64.0**	58.2
Proliferation of weapons of mass destruction	**62.7**	61.5

Examining the sample as a whole, we see that the status – military or civilian – of the respondent has an influence on the answers, as was expected. However, the proportional differences, reported in Table 4, are not large on average except for a few types of threat like mass immigration from foreign countries (a gradient of nearly 11 percentage points for the cadets) and the possibility of being involved in a conflict in the Middle East (a difference of nearly 10 percentage points). For seven threats out of nine it is the young cadets that show the most concern, while for two others (i.e., threats to the environment and the possibility of a conflict over the control of raw materials) it is the university students that are most concerned. This is an average datum: more significant differences between the perception of threats by the military cadets and their civilian peers are seen for countries like France and, to a lesser degree, Switzerland and Sweden. Among the students in the various disciplines the average datum shows greater concern by economics students than by their peers in other faculties.

In a breakdown by gender (see Table 5), women are most worried about likely threats to national security on average. In particular, they display a higher percentage of concern for seven threats out of nine, but the gradient is sizeable (nearly 10 percentage points) only for threats to the natural environment. Men, on the other hand, are significantly more concerned about illegal immigration from other countries.

Table 5. Perceived threats by gender (%)

	Male	Female
Armed conflict over the control of vital raw materials	60.9	**61.2**
Mass immigration from foreign countries	**57.8**	51.6
Terrorism	63.2	**66.5**
International drug trafficking	74.0	**77.5**
Organized crime	79.3	**79.9**
Environmental problems (air pollution, etc.)	60.0	**69.8**
An armed conflict in the Middle East	**51.1**	49.9
Attacks on computer networks	64.0	**65.1**
Proliferation of weapons of mass destruction	61.3	**63.7**

But how much does the level of information received on military security issues influence the perception of threats? The responses given to a question on the level of information provided by the media on military and security issues allow us to take a look at this aspect as well. It should be said at the outset that the respondents average evaluation of the level of information their country's media provide on these issues is not very good. Indeed, the sample is divided into 45.7% who consider it good and 48.3% who consider it poor (the remaining respondents have no opinion). The data, especially the information received regarding threat perception, show by trend that those most informed tend to perceive the individual threats more distinctly, with the interesting exception of terrorism: it seems here that this type of threat is now globally and universally apprehended, independent of how informed one is in this regard.

Table 6. Level of information by likelihood of single military threats (%).

Threat likelihood	How do you judge the level of information that the media give you on military issues?			
	Very good	Somewhat good	Not good	No opinion
Mass immigration	61.0	54.3	56.4	54.3
Terrorism	63.0	60.3	69.4	51.4
Middle East armed conflict	59.0	46.0	55.4	38.5
Proliferation weapons mass destruction	67.2	60.3	63.2	61.0

Not all the interviewed young people show the same level of interest in national security issues, however. Naturally, the cadets are more interested than their university peers, the male respondents are more interested than the females, while the students of political sciences show the highest level of interest among those attending the various university faculties in general. This aspect, which I only touch on for reasons of brevity, is significant because there is a strong positive correlation between the level of interest and the level of threat perception, as exemplified by three threats in Figure 4. This seems to mean that where perception is low it may be the result of little interest in the problem and also, as we saw earlier, of a lack of information on security issues.

Figure 4. Likelihood assigned to threats (in %) in relation to interest in security issues

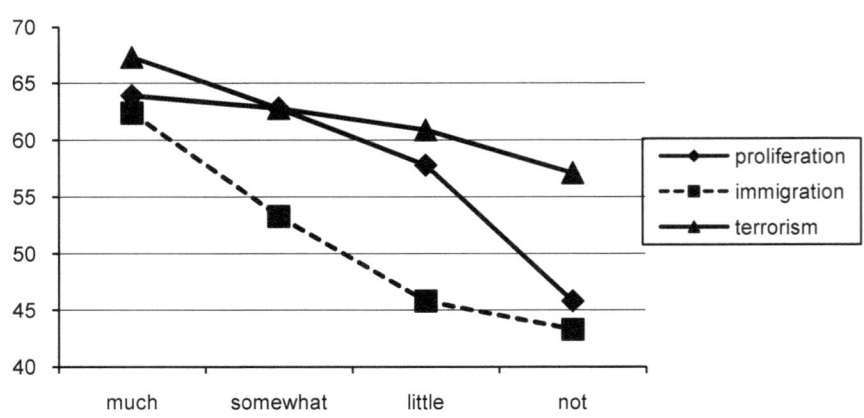

The same type of positive relationship is seen between the level of threat perception and the willingness to accept an increase in the national defense budget to address growing security threats. One significant datum regarding the entire sample is that a 64% majority of the future elites examined think that raising the defense budget to some extent is a necessity. This datum naturally differs from country to country and it may be somewhat interesting to present the national differences graphically (Figure 5). This illustrates that in some countries (e.g., Slovenia, Spain, Switzerland, Turkey) agreement with enhanced defense spending is the minority view.

Figure 5. Agreement to raise defense budgets by country

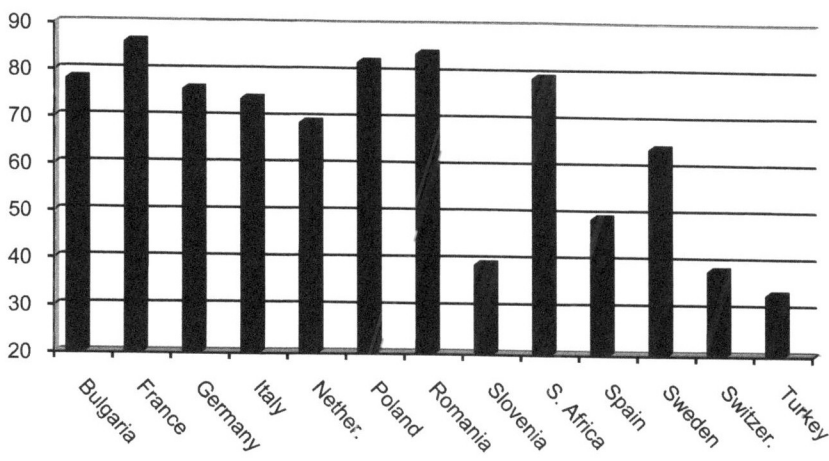

The change in threat perception over time has been analyzed through a comparison between the interviewed young people in their first year of studies and those in later years, substituting a diachronic analysis with a synchronic one and assuming that the variations during attendance of the academic courses can give a fair approximation of a four-year variation. The data are represented graphically in Figure 6 for greater visual impact. They tell us that – imagining an ideal regression line – the threat perception generally increases over the course of the years, with the exceptions of the threats represented by mass immigration (in decline) and those pertaining to a Middle East conflict and an attack on computer networks, which remain more or less stable.

215

The analysis of the perceived threats also includes an assessment of who should primarily be responsible for addressing the threats and protecting the population from them (Table 7). Today, the responsibility and the initiative to resolve conflict situations with armed interventions is for the most part entrusted either to supranational bodies or to the current hegemonic power, the United States. The interviewees were therefore asked to indicate the degree to which they were in agreement with entrusting such an initiative to one of these entities. The sample responses overall clearly privilege the United Nations, even if not with the near-unanimous agreement that one might expect, followed by NATO and the European Union, with the percentages reported below. As can be seen, the possibility of giving the U.S. freedom to intervene met with almost total disagreement.

Figure 6. Likelihood assigned to threats (in %) by course year.

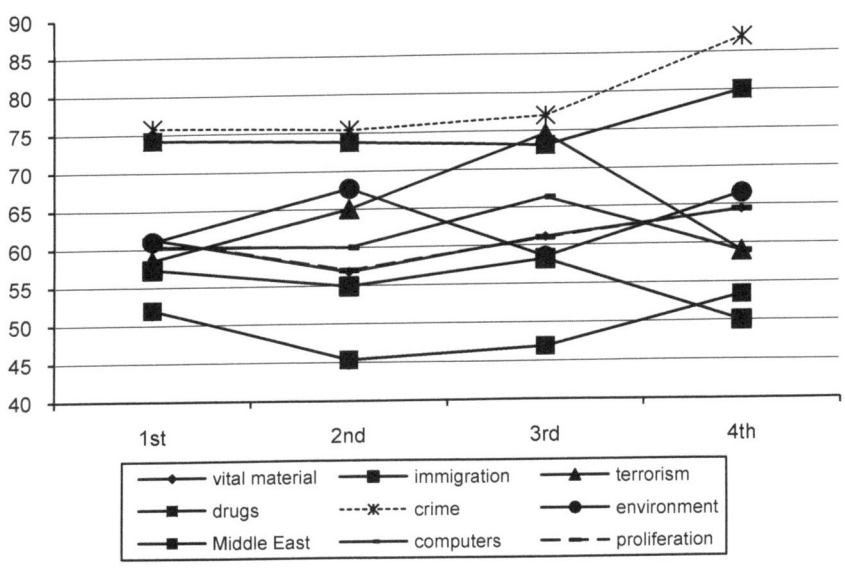

Table 7. Approval of intervention depending on political entities

Should the following entities have freedom to intervene on their own initiative?	
United Nations	58.4%
NATO	33.9%
EU	22.7%
USA	12.4%

Approval for the individual entities is of course different from country to country. Since these differences are significant they are graphed in Figure 7. From the figure we can see that the future elites of a group of countries (Slovenia, Switzerland, South Africa, and Spain) appear to be more oriented towards U.N. leadership and control over military actions, while another one consisting of France, Poland, and Turkey has more reservations towards that organization. NATO appears to enjoy particularly strong support among the interviewees from Germany, the Netherlands, and South Africa, while the European Union garners average support from the respondents of all the member countries except for Spain. Sweden, Switzerland and Turkey are obviously less supportive of the possibility of intervention by the EU. Among the countries that disagree with U.S. military initiatives, the Netherlands and South Africa are the ones most open to this possibility. There are no significant percentage gaps between the university students' and the cadets' opinion on this issue, nor is there any strong disparity due to gender.

Figure 7. Approval of interventions (in %) by political entities and countries

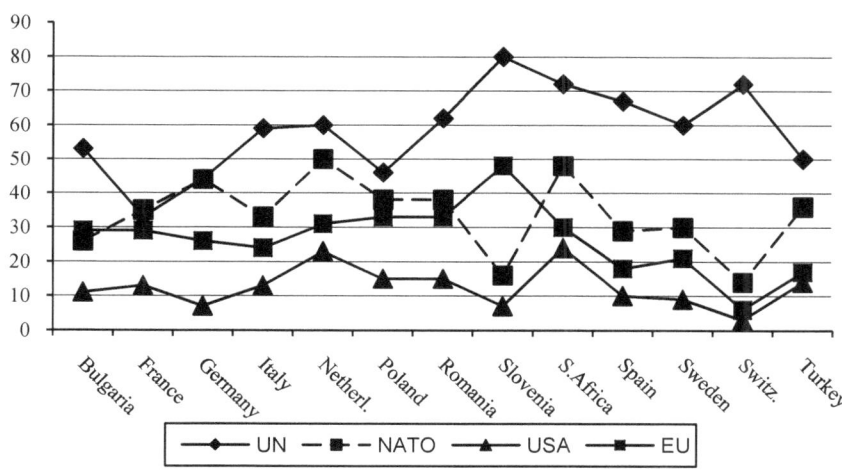

Discussion

From the data presented in the preceding section we can see that the future elites of the examined countries consider a significant set of threats to have lost currency. Among them is the possible involvement of one's country in internal or external conflicts of Third World countries (with the significant exception of a Middle East conflict), as well as nuclear threat, whether deliberate or accidental. A moderate level of likelihood is also assigned to armed aggression by another country. The most deeply felt threats are equally divided, as mentioned, between threats that challenge the police and threats of a military nature, with obvious mingling and overlapping between the two areas.

The diverging importance given by the response groups of the individual countries concerned by the individual threats examined (the most significant ones in terms of assigned likelihood) depends on several kinds of motivation. In some cases there is a clear motivation due to geographic position, for instance regarding mass immigration, where the elites of very exposed countries like Italy and South Africa show the greatest concern. In other cases we can point to situational motivations. The fear of attacks on computer networks, e.g., is particularly felt by the future elites of those countries where the development of (and the dependence on) computer networks is most extensive and far-reaching. The greater probability assigned to the criminality/drug complex by the interviewees from the former communist countries may stem from the fact that this problem actually appear to be dire and more pressing there. But in some cases one can spot motivations of a third type, which I would term emotional. This applies in the case of the Dutch respondents who declare a strong perception of threat produced by uncontrolled mass immigration, even though the Netherlands is a country that is not particularly affected by this phenomenon, or in the case of the Romanians, who display strong sensitivity to the terrorist threat even though their country has been much less affected than others by that type of menace.

Breaking down the numbers, it is interesting that no large gap in threat perception between cadets and university students is evident: the gap is only significant for two items, i.e., mass immigrations and the possibility of being drawn into a conflict in the Middle East. The lack of such a significant gap, which one might have expected, suggests that the future elites are beginning to have a fairly shared and probably realistic view, independent of their training processes and career choices; not least since the differences are also minimal between the students of the different faculties.

In the analysis by gender, distinct percentage gaps are not shown either, with the exception of women's greater concern for environmental issues (this may be due to the ancient feminine depiction of *mother earth*). Overall women are more concerned about threats percentage-wise than men, but this seems consistent with an evolutionary psychological view of female nature.

A positive relation also exists between the intensity of threat perception and the level of information on issues of national military security and interest therein. The same type of positive relation is also seen between the level of threat perception and the willingness to accept an increase in the national defense budget to address growing security threats. Moreover, the perception of these threats seems to increase with the years of study: the third- and fourth-year students display, on average, higher percentages of concern than those just beginning their studies. The average young person concerned about the likelihood of the assessed individual threats is that of a person who feels sufficiently informed by the media, who takes more interest than others in military security issues and, consequently, feels more willing to accept economic sacrifices to ensure an adequate defence against those threats. This position seems to consolidate as the young person advances in his or her studies respectively age.

The overall picture of the examined future elite's perceptions can be completed by an analysis of their attitudes towards the major international players they wish – or do not wish – to entrust with the defense against (and/or prevention of) these threats. This analysis is made on the basis of the freedom for armed intervention – a significant although not exhaustive element – that the young interviewees are willing to grant to these international players. Judging from the present analysis, the U.N. is the entity to which the respondents are most willing to entrust the security function. This majority support does not approach unanimity, however: evidently reservations (which have not been investigated) exist with regard to how the U.N. works. This explains the support in favour of NATO as an entity that should be granted the freedom of intervention; support that would not make much sense if confidence in the U.N. were substantial.

What is clearly refused by nearly all of the interviewees is the notion of assigning this power and function to a single country, in this case the hegemonic power. Only a tiny minority wants the United States as the world's policeman. Behind the U.N. and NATO, the EU ranks third as a potential international player with a military role, and it appears that this acceptance is growing.

Bibliography

Bauman, Z. (2002). *Society under Siege*. Cambridge: Polity Press.
Beck, U. (2000). *The Risk Society and Beyond. Critical Issues for Social Theory*. London: SAGE.
Caforio, G. (Ed.) (1998). *The European Cadet: Professional Socialisation in Military Academies*. Baden-Baden: Nomos.
Caforio, G. (Ed.) (2007). *Cultural Differences between the Military and Parent Society in Democratic Countries*. Amsterdam, New York: Elsevier.
Fukuyama, F. (1989). The End of History. *The National Interest, 16*, 3-18.

Giddens, A. (1991). *Modernity and Self-Identity. Self and Society in the late Modern Age*. Cambridge: Polity Press.
Maniscalco, M. L. (2004) *Opinione Pubblica, Sicurezza e Difesa in Europa*. Soveria Mannelli: Rubbettino Editore.
Mearsheimer, J. (1990). Back to the Future. Instability in Europe after the Cold War. *International Security, 15*(1), 5- 56.
Zanetti, A. (Ed.) (2004). Società del Rischio, Pace e Sicurezza. In M. L. Maniscalco (Ed.), *Opinione Pubblica, Sicurezza e Difesa in Europa* (pp. 55-93). Soveria Mannelli: Rubbettino Editore.

Audrone Petrauskaite

Civic Values in the Context of the Military Profession: The Lithuanian Case

It is obvious that the objectives of military education are based on the requirements of the society. Lithuanian society is creating a national state in the era of globalization, transnationalization, and internationalization. Currently the issues of national identity and the purposes of national security policy are becoming very important, as are the issues of international cooperation, the creation of the civic society, and the objectives of international security. From this point of view, the value system of both Lithuanian society and military has become a current issue in military education.

The goals and tasks of military education in Lithuania have always been determined by Lithuanian political aims and the need for social development of the country. 15 years ago, the military academy of Lithuania was established, combining professional military studies with other academic branches. This was an expensive but rather necessary and important decision, which met the requirements put forward by society: Lithuanians wanted to see well-educated and intelligent persons ready to fulfill military tasks for the sake of national defense and security as officers of the Lithuanian Armed Forces. The society aimed at carrying out the most important task to defend the state believing in high morals and deep national conscience of the military.

Recent transformations in the Lithuanian national defense system have brought some confusion both to society and the military. According to new military missions and tasks, the Lithuanian Armed Forces are introducing changes to their structure with the aim to form professional military troops: small, well-trained, very mobile units, which are ready to participate in different international military operations. These transformations have created some reservations in the minds of Lithuanian people regarding the necessity of the national armed forces and their goals. According to Lithuanian tradition, the army has always been "the second biggest school of the nation", where national traditions, historical memory, and the highest national and civic values are cherished (Raštikis, 1957). The main and greatest duty of the Lithuanian Armed Forces has always been to defend the native country and its people from potential enemies. Currently, our society has some doubts: The participation in international military missions far away from the homeland has little in common with national defense tasks. What is more vital at the moment is that Lithuanians think that the professionalization of the armed forces will result in changes in the motivation of the military: Soldiers will gradually become hired guns and military service will lose its patriotic meaning. So the changes of the military personnel's organization will have effects on the transformation both in the value system and the moral orientation of the military.

Under these circumstances, the following question arises: *Could the process of transformation, globalization, and internationalization threaten soldiers' national and civic identity?*

This question is very complex. From my point of view, in order to answer this question we need to determine:

a) What values constitute the professional soldiers' identity?
b) What is the relationship between universal, national, and democratic values in the civic conscience of a person?

According to E. R. Micewski (2005), a postmodern paradigm of armed forces has started taking shape since the end of the Cold War and has brought about major changes to the military ethos and nature. It has changed the social status of the military and also its link with the civilian society. The military is losing some of its isolation and exclusive position: it has become open to the society, and more importantly, it has become an integral part of it. As a result, society now reflects all military views, values, attitudes, and problems. The military is becoming more professional and military service is turning into a profession with its professional moral standards and value system. Moreover, a soldier, as a representative of his profession, is given the chance to be a human being like other professionals. In my mind this is the most important and positive transformation in the conscience of the military. Of course, the soldier has always been a human being in the first place, but historically, the military tended to forget about this. When I entered the military the first rule I had to learn was: "I am primarily a military servant and only in the second place a human being".

In my opinion this tradition was exclusive to the military domain. Has anyone ever heard anybody saying, "I am primarily a teacher and only in the second place a human being"? But why does this notion only occur in the military? All professionals have moral standards to keep in their professional activities. All professionals have to forget about their human weaknesses and fears when they seek to carry out their professional duties as perfectly as possible. This requirement is indispensable if one wishes to adequately serve one's people. On the other hand, all codes of professional ethics are based on universal human values, i.e., respect, humanism, tolerance, responsibility, etc. And professional honor in its essence means human sensibility and personal responsibility for the better implementation of professional tasks. From this point of view, military service as a profession is not the exception to the rule (Gabriel, 1987). The soldier must perform his professional duties according to the requirements of a military code of honor, which includes crucial human principles and moral norms. Practically every soldier has his personal system of professional values, which are based on professional ethical principles, moral norms of the society, and his personal attitude toward his professional activity (intellectual and emotional motivation, knowledge, life experience, etc.).

Both as professionals and human beings, soldiers are members of society and they form their value system under the influence of many factors (e.g., social, political, economical, and psychological ones). Moreover, they are citizens, meaning that their personal moral maturity is closely related to their civic and national value system. So the soldiers' professional ethical virtues and attitudes are formed under the influence of their national and civic consciousness, making this one of the most important factors in the professional moral conscience of soldiers.

The civic conscience of a person is characterized by his identity, civic virtues, political activity, and social demands for civic efficiency. According to T. H. McLaughlin (1994), mature civic society cannot be characterized formally (i.e., by formal attributes of a citizen, his formal rights, formal guarantees for his social activities, and by the minimalist interpretation of civic virtues like "here and now"). The introduction of crucial civic values, e.g., civic identity and virtues, has to be extensively implemented into the conscience of society.

Today, the civic identity of a person necessarily combines a wide spectrum of sociocultural and psychological aspects. In this sense, civic values include among other things freedom, equality, solidarity, and tolerance, all of which could characterize postmodern societies as being open-minded, multicultural, multiconfessional, and as manifesting mutual understanding. Therefore, these values are not contrary to such central national values as the nation, national language, national culture, national historical memory, and traditions. From this point of view, civic identity is very closely related to national identity. Political activities of citizens have to embrace all democratic processes of the society. If a given society regards a citizen's zeal as a civic value, then this reflects the civic maturity of that society as a whole. On the other hand, an interaction of elements (social, political, economical, and cultural) has to be provided in favor of a successful and effective activity of citizens.

The last and the most important factor in the formation of postmodern civic societies is the issue of civic virtues. Basic civic virtues like respect, responsibility, tolerance, justice, loyalty, etc., have to exceed their limits and aim for universal human values, for their deep understanding and perception. Without any doubt, the civic conscience of a human being is formed on the basis of civic, democratic, national, and universal human values.

So the conscience of soldiers is a convergence of their moral, national, and civic values. All these values can be seen to form the basis for the right moral attitude towards the professional activities of a military person. Theoretically, the professional maturity of a soldier has to be similar (or equal) to appropriate moral and civic values of society. We can compare the test results for the civic conscience of Lithuanian society and the Lithuanian military and see the difference.[1] The major part of the respondents answered that Lithuania is *"The state where I was born and live"* (83% of the servicemen and 64% of the civilians). Accordingly, the majority of the soldiers (61%) as well as a considerable percentage of the civilians (30%) think that Lithuania is *"My native place where I was born and live"*. And only 24% of the soldiers and 25% of the civilians believe that Lithuania is *"The territory where Lithuanians are living"*. As could be noted, these characteristics of national and civic identification have almost the same tendencies in both groups of respondents:

- The major part of the respondents identify with the Lithuanian state
- The home town or village is very important for both groups of respondents
- The Lithuanian territory is not so important for the respondents

The results of these tests allow us to conclude that the national state is the basic civic value of the Lithuanians. But there are also some other indicators as to the identification of Lithuanians. The respondents' value system is as follows (see Figure 1):

Figure 1. The most important national values for Lithuanians (in %)

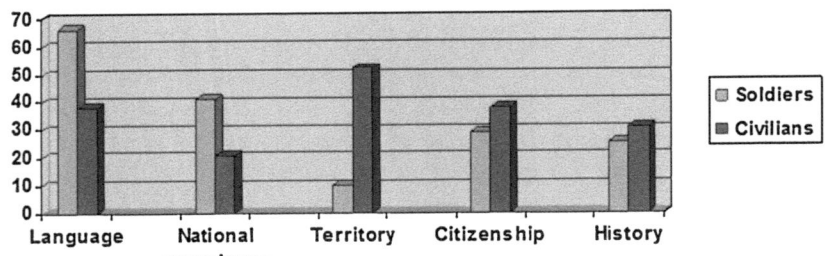

---

1   The results of the tests are cited from: Matulionis, A. (2007). *National Identity of Lithuanians.* The Institute for Social Research; and Petrauskaite, A, Kazlauskaite, R. & Zigaras, F. (2007). *Civic Education in Lithuanian Military Forces: Development, Experience, Issues and Future Solutions.* Vilnius: General Jonas Zemaitis Military Academy of Lithuania.

- The most important values are: national language, Lithuanian territory, Lithuanian citizenship, national conscience, and history;
- There are some differences between the answers of soldiers and civilians. The most important values for the former group are language and national conscience, for the latter group it is the territory of Lithuania, its language, and citizenship.

These results seem to confirm that the civic identity of Lithuanians is based on national values. Moreover, they suggest that soldiers are maybe more oriented towards national values and more motivated by patriotism than the civilians. Then again, there is reason to doubt the national maturity of the respondents based on further results from the test (see Figure 2).

Figure 2. The best symbols of Lithuania (in %)

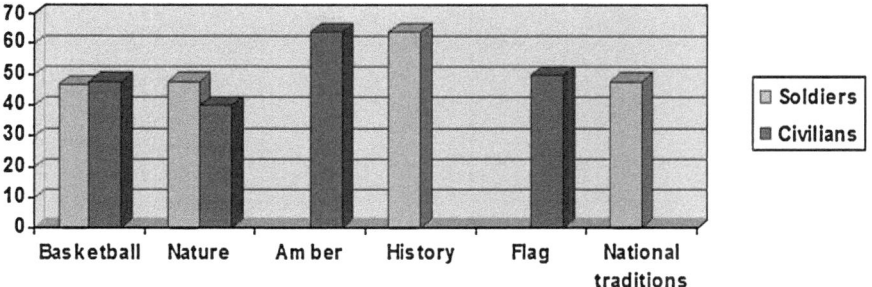

Judging from the test results it is obvious that amber and the state flag are the greatest pride of the civilians, while history and national traditions are the country's most honored symbols in the eyes of the soldiers. Evidently, the question "what is the best symbol of Lithuania?" reveals that there are quite different positions among civilians and the military. Thus it is very difficult to demonstrate which respondent group is more motivated by national or by civic attitudes. On the other hand, both respondent groups are quite proud of Lithuanian basketball and the country's nature. This point of view is based on a widespread mental pattern. National values such as "favorite sport" (basketball, football, ice-hockey, etc.) or "native country's nature", or "national women's beauty" are typical "national" symbols in many countries as they reflect the national identity of the people. Actually these values are an interesting and rather complex phenomenon of national mentality. In my opinion it can be attributed to global or mass culture, and not to national or civic identity. These latter values express the ideals and expectations of the nation and belong to an emotional, not a rational human mentality.

Of course, the concept of values implies social attitudes that govern citizens' personal behavior (Kuzmickas, 2001). Emotions play an important role in various individual and social activities, but the national and civic identity of a person or society should have a greater importance than the victory of the national basketball team or the ecological situation in a certain region. Currently, civic and national values of civilians and the military don't differ much and this is characteristic of postmodern armed forces. On the other hand, both respondent groups reveal a minimalist interpretation of civic values and national identity by choosing emotional motivations over rational ones. In doing so, they contradict the notion of "real" maturity of a civic society.

Other test results of the soldiers show that there is some disagreement between the civic and professional values in their conscience. Important civic and moral virtues such as responsibility, justice, and honesty are priorities in the soldiers' system of values. On the other hand, tolerance and friendship are valuable for soldiers as humans but not as servicemen (see Table 1). These results could lead to the conclusion that the traditional way of thinking is very strong in the military: they have a very strong sense of military virtues, which stand in a certain contrast with universal human virtues.

Table 1. Importance of civic and military values for soldiers.

Valuable feature	For civilians (%)	For the military (%)
Justice	67	67
Responsibility	65	64
Friendliness	63	9
Tolerance	51	20
Intelligence	49	13
Honesty	47	75

During the long history of the military profession, an awareness of a certain uniqueness and exclusiveness has been introduced into military ethics and the professional value system. Justice, responsibility, honesty, courage, and duty as traditional professional values still play the most important roles in shaping the identity of military professionals. On the other hand, these values cannot be opposed to universal human values. The system of personal values is formed on a complex basis (see Figure 3). The principles of the professional value system are deeply rooted in national and civic values. And all the above-mentioned values are born from universal human values. It is evident that the personal value system of the military has to include such basic human values as respect, tolerance, and humanity in the first place. It is a requirement of the postmodern society and an educational task for the future military.

Figure 3. Value system pyramid

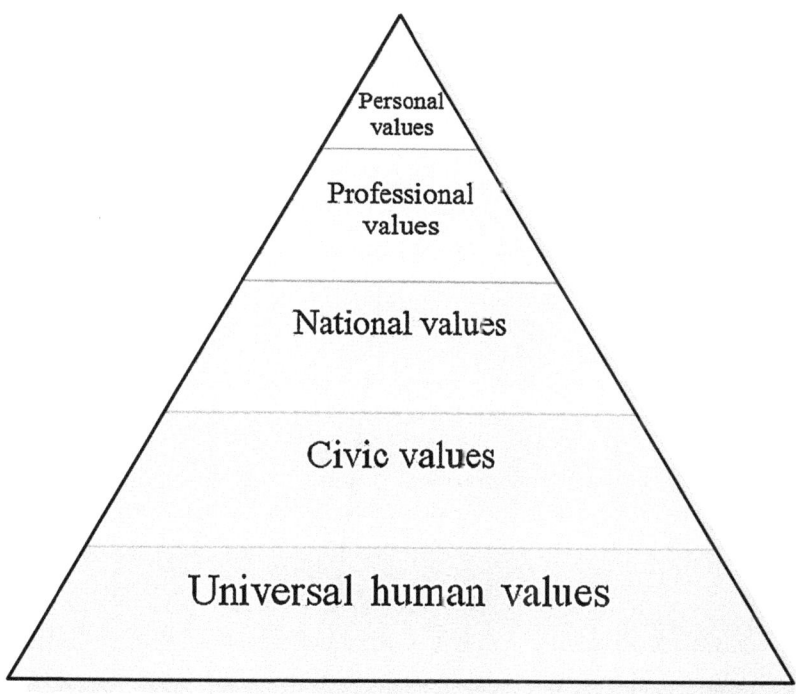

Conclusions

- According to the test results, professional and national values represent priorities in the professional conscience of the Lithuanian soldiers. This stands in conflict with the postmodern paradigm of military forces and the requirements of the civic society. The above-mentioned indicates the necessity of civilian education for the military.
- A deep professional and patriotic esteem of the national state inherent to the military may hinder the development of predominantly materialistic motivations for professional military service.

- A civilian education of the military needs to be established, including universal human, civic, national, and traditional professional values.
- Moreover, the goals of civic education have to engage society's attention. This is obviously a task for both the educational and military systems of Lithuania.
- The military needs to turn its focus back to national defense and convey this message to the institutions of civilian education, especially schools, colleges, and universities. This could help to develop the Lithuanian youth's patriotism and to attract individuals to pursue a professional military career.

Bibliography

Gabriel, R. (1987). *To Serve With Honor*. New York: Praeger.

Kuzmickas, B. (2001). *Laimė, asmenybė, vertybės* [Happiness, Personality, Values]. Vilnius: Lietuvos Teisės Universitetas.

McLaughlin, T. (1994). *Contemporary Philosophy of Education: Democracy, Values, Diversity*. Kaunas: Technology.

Micewski, E. (2005). Leadership Responsibility in Postmodern Armed Forces. In E. Micewski & D. Pfarr (Eds.), *Civil-Military Aspects of Military Ethics, Volume 2* (pp. 5-12). Vienna: National Defence Academy Printing Office.

Raštikis, S. (1957). *Kovose dėl Lietuvos* [The Fight for Lithuania], *Volume 2*. Los Angeles: Lietuviu Dienos.

Asa Kasher

Military Ethics between Code and Conduct

The Problem

The gap between what is done and what should have been done is ubiquitous. One becomes aware of it when first scolded by a parent for doing what should not have been done. When one is a pupil, a student, or a cadet one knows that any manifestation of a gap in one's activity might be followed by one's being reprimanded by some instructor. Every sermon one happens to hear, whether in a religious temple or a political rally, reminds one of some gap between ideals that should have governed major aspects of human life and reality, which is commonly far from fully manifesting those ideals.

The present paper focuses on the problem of gaps between ideals of proper behavior of combatants and their actual behavior under various circumstances of military activity. The problem of such gaps has taken a new shape with the emergence of the "strategic sergeant", a combatant who does not have many troops as subordinates, who does not bear any heavy formal responsibility but who knows that a single wrong action on his part might result in a new strategic situation or at least in some grave problem for the government of his state on a level of international relations.[1]

Our study of "the ethical gap" between code and conduct in military activity is theoretical for more than one reason. An obvious reason is that philosophical theories can help us both in understanding the ethical gap and in enhancing attempts to bridge it or at least narrow it. Another reason is less obvious, but is related to an aspect of military ethics commonly ignored. The seemingly simplest and most natural question to ask about the implementation of a certain articulated conception of military ethics in a military force is how wide the ethical gap is. As a matter of fact, usually we do not have a real answer to that question. One can have some personal experience within the framework of a military force related to the ethical gap. One can have access to reports of certain investigations, whether legal or professional, of certain events or phenomena. One can consume apparently related media reports. However, these sources of information, even if taken to be reliable, do not provide one with a general answer about the width and depth of the gap.

---

1   For the sake of generality, we are not going to make assumptions about the form of military ethics used for describing the ideal that should govern military activity, be it values, principles or virtues.

To be sure, this is not an observation about military ethics in particular. Parallel observations can be made about medical ethics, teaching ethics, business ethics, and what have you. Some attempts to develop tools for evaluating such ethical gaps have been made. An example in military ethics is included in the U.S. Army FM 22-100 (1990), where each presentation of a leadership value is accompanied by a very detailed list of questions that enable a commander to determine to what extent the value has been successfully implemented in one's unit. Another example is the recent Integrated Ethics project of the U.S. Department of Veteran Affairs (VA), where the identification of ethical gaps is followed by a systematic effort to understand and then eliminate them.[2] However, even such attempts do not pretend to provide the user of the suggested tools with an accurate evaluation of the width of the ethical gap in general or of some specific ethical gaps in particular.[3]

In the following sections, ethical gaps are shown to emerge from certain independent problematic circumstances. The ethical gaps we are going to discuss are related to various aspects of military ethics, not only moral ones.

Knowledge

The simplest ethical gaps involve gaps in required knowledge. No professional activity can be properly performed in a certain context without knowledge required for proper planning and proper performance. Hence, ordering a military unit to accomplish a mission without providing it with the required intelligence is more often than not an ethical gap.

One example appears in a description given by a Canadian Battalion Commander, Lt. Col. Ian Hope, of his experience in Afghanistan (2007). Under his command, Task Force Orion was at one point ordered to retake from the Taliban the district centers in Nawa and Garmser by 4:30pm the next day. When asked by his brigade commander, "Are there any questions?" his reply was, "Roger, Sir. Just one. Where are Nawa and Garmser?". "They are in Southern Helmand", answered the brigade commander. "Roger", replied Hope, "but we have no maps."

---

2 See U.S. Department of Veteran Affairs, National Centre for Ethics in Health Care. Available: http://www.ethics.va.gov/ethics/integralethics.

3 See also the questionnaire used by the Ethisphere Institute in the recent Government Contractor Ethics Ranking, based on U.S.A. Federal Acquisition Regulation. The components of the ranking are Code of Ethics and Business Conduct, Leadership and Tone from the Top, Internal Control Systems and Ethics Training and Communication program. Evaluation of actual behavior is not among the measured ingredients of the Ethics Quotient. Available: http://ethisphere.com/wp-content/uploads/2008/04/wme-eq-questionnaire_pub_0003.pdf.

Another example involves soldiers at checkpoints, whose mission is to capture people with terrorist intentions. Obviously, without ROEs that enable such soldiers to identify suspects, they are bound to commit errors, even fatal ones.

Notice that the ethical gap is not necessarily manifest in the ensuing behavior of Hope's battalion or the soldiers at the checkpoint. The ethical gap resides in the preceding violation of principles of professionalism in military activity. However, the preceding ethical gaps can lead to additional ethical gaps, such as causing casualties or collateral damage which could have been avoided.

Understanding

Much of what we encounter that constitutes ethical gaps results from neither wickedness nor negligence but rather from a lack of understanding or even from misunderstanding of major parts of the framework within which people act.

To understand what one does as a person of a certain profession, duty, and responsibility, or what one is doing under certain circumstances, given certain mission and authority, is first of all to have in mind a broader picture of the framework of one's activity. Such a broad picture renders one's activity meaningful. It provides one with adequate answers to questions such as "what is the meaning of my present mission?", "what is the meaning of our military presence in a certain state, such as Iraq or Afghanistan?" or "what is the meaning of our profession, within the framework of a constitutional democracy?".

The importance of understanding one's activity, in the sense of knowing the meaning of it, is clearly seen when a lack of understanding takes place, causing the emergence of an ethical gap. During the Second Lebanon War, there were some instances of commanders not even trying to accomplish missions they had been given, since on the preceding day they had received a series of missions to accomplish, each of which was eventually cancelled. Not trying to accomplish one's mission is indeed an ethical gap. We take it for granted that those commanders were professional, order abiding, and courageous. Why, then, were they tempted to produce ethical gaps of such significance? Our answer is that they did not understand the missions they had been given and not being able to embed them in broader pictures of the ongoing operation or war, they embedded them in their own personal experience of missions given and cancelled. The first principle on the IDF formal list of the principles of warfare is "Perseverance in making efforts to accomplish the mission against the background of the goal", the latter being the broader picture in which the mission is embedded and becomes meaningful.[4]

---

[4] The importance of understanding and the influence it has on behavior can be shown in many areas and ways. For example, it has been shown that telling smokers their lungs age significantly

Full understanding of an action involves more than being aware of a broader picture within which the action becomes meaningful. When an action is carried out by members of a certain profession, an adequate understanding is required of the nature of that profession, which has to be manifest in every action taken within its framework. Similarly, when an action is carried out by members of a certain agency, adequate understanding is required of the nature of that agency.

Understanding the nature of a profession or an organization requires knowledge of its identity. Understanding the profession of military command requires the ability to answer the fundamental question of its identity, namely, "what is a military commander?". Again, lack of understanding of professional identity and organizational identity, and all the more a misunderstanding of such identities is bound to lead to ethical gaps. Here are some examples, couched in terms of values, which are commonly used in depictions of a professional or an organizational identity.

Some military values are related to the goals of military activity in general. In a democratic setting, the goal of military activity in an international conflict that involves combat is self-defense.[5] Thus, e.g., when pilots or soldiers target a person, reliably marked as a terrorist, they ought to know that the purpose of the targeting is preemptive self-defense. If they take the purpose to be punishment, they might show less perseverance. If they take the purpose to be retaliation or deterrence, they might be less sensitive to collateral damage. The moral and ethical gaps of targeting killing as a form of punishment, retaliation or deterrence will become even deeper when such additional ethical gaps emerge (Kasher & Yadlin, 2005).

Some military values are related not to the goals of military activity but to the means usually chosen in their service (see Kasher, 2008). One obvious example is that of discipline. The obvious manifestation of discipline is obedience. Every case of disobeying legal orders[6] is taken to be a blatant breach of discipline, a legal gap. Quite often, facing a case of such disobedience, people leave it at that; it is regarded as a legal gap that is to be dealt with by law enforcement units or bodies. In our view, however, to consider every case of disobedience solely as a

---

    improves the likelihood of them quitting smoking (see Parks, Greenhalgh, Griffin & Dent, 2008). Since smoking is an intricate habit, understanding the role it plays in the broader picture of lung aging is not sufficient in order to reach a decision to quit smoking, but a step of embedding one's smoking in a broader picture has been shown to be of some significance.

5   Military forces have recently been deployed in fighting terrorism (which is usually also a case of self-defense) as well as in peace keeping and in humanitarian intervention, where the goals are different. Our present discussion pertains to the former uses of military force.

6   In some states, including Israel, disobeying an illegal command which is not manifestly illegal is also a breach of discipline as well as illegal.

problem of a legal gap and to be treated, prevented, prosecuted, or punished as an instance of a criminal activity is a mistake. Disobedience is always also an ethical gap which should be treated as an indication of some ethical weakness. In this context, again, understanding is required: this time an understanding of the value of discipline, which is part of an adequate understanding of the identity of the military force. Obedience is just an aspect of discipline. The former simply involves a relationship between orders and performance that can be described and treated on a legal level, but the latter involves a relationship between persons, commanders, and their subordinates. "The way of leadership", said Zhuge Liang, the second century Chinese general, "puts education and direction before punishment" (Zhuge & Ji, 2005), or in other words, ethics before law (Kasher, 2002).

The relationship of discipline involves first and foremost mutual trust, which cannot be treated on any legal level, but only on the educational levels of ethics. Hence, a case of disobedience is an indication of a problem on the level of mutual trust between commander and subordinate. Such a problem is also an ethical gap that should be appropriately bridged.[7]

An additional type of ethical gaps caused by a lack of understanding and even misunderstanding is related to a major ingredient of being a professional, namely being able to justify one's professional activities and answer "why?" questions about them. In the mid-1990s I realized that some IDF troops, who at the time served in Southern Lebanon, thought that according to their ROEs they should target a comrade if he had been kidnapped and they had failed to rescue him from the hands of the kidnappers. This was indeed a major ethical gap, caused by ROEs which had not been explained, and therefore were not understood. To make things worse, eventually, they were being misunderstood. Those ROEs used the expression "by all means" when they ordered the soldiers to try to stop the kidnapping and required that the soldiers shoot at the kidnappers, thus jeopardizing the life of the kidnapped comrade whose fate rested with the kidnappers. They failed to understand that they might put the life of a comrade at risk while trying to rescue him. They also failed to understand that the expression "by all means" did not imply a total exemption from adherence to military values such as comradeship. Fortunately, no casualty was caused by that ethical gap. Later on, the ROEs were changed so as to specify their goal, i.e., rescuing the kidnapped soldier. Notice again that the roots of one ethical gap can reside in a previous one, in the actions of commanders who do not properly explain ROEs

---

7   The Inquiry Commission that investigated the Second Lebanon War emphasized problems of disobedience that had been encountered during the war. Our discussion leads to the conclusion that deeper problems should be investigated, namely problems of lack of trust or even of mistrust of commanders and their own commanders.

and thereby leave ample room for a lack of understanding and for misunderstanding.

A related issue that should be addressed is that of accountability. A Google search for "accountability" (done on April 15, 2008) resulted in a list of 26,800,000 hits. Some of these websites assume that accountability is synonymous with transparency and for obvious reasons the media reinforce such a narrow meaning of "accountability". A person who thinks that in one's professional activity one should manifest the ethical value of accountability is bound to act improperly if "accountability" is taken to mean transparency. Misconduct is bound to take place under such circumstances, since the point of view of members of the public, to whom one might feel to be accountable, will under such circumstances play an exaggerated role in shaping one's behavior.

The major component of accountability is not transparency but rather professional or institutional justifiability of one's actions and omissions. To be accountable means to be able and willing to justify one's activity to related stakeholders. Justification in a professional setting rests first and foremost on the foundations of professional knowledge, independent of the point of view of members of the public. An officer is held accountable to his commanders in the sense that he owes them a justification of his actions and omissions during his military activity. Such a justification ought to be professional (or institutional, or both) rather than popular. Accountability is thus closer related to responsibility and integrity than to transparency (Stein, 2003).[8]

Moral Responsibility

Let us now consider the case of a soldier in charge of a checkpoint. The mission is to deter or even capture persons who are insurgents, terrorists or both. In order to accomplish his mission, the soldier has to check everyone who is interested in crossing the line from one side of the checkpoint to the other. Checking a person can be done effectively and politely, but every now and then it involves unnecessary humiliation of some people. Such occasions are, undoubtedly, ethical gaps.

Two explanations are often given for such instances of military misconduct. We would like to show that both are wrong. The causes of such ethical gaps are deeper and more pernicious than what meets the eye.

---

[8] See also The Charter of Accountability, which takes the requirement of accountability to involve required explanation and justification of action and omissions to stakeholders. The charter requires explanation and justification of intentions, too. We assume that not intentions in a broad sense are required but rather practical plans. Available: www.freedomtocare.org/page313.htm.

One explanation has taken the form of a slogan: "Occupation corrupts". Let us disregard political motivations for common usages of the slogan and take it, for the sake of the argument, to be a genuine attempt to explain the encountered ethical gaps. To start seeing what is wrong with such an attempted explanation, let us consider an analogy. Assume that drivers show a much stronger inclination to behave improperly than pedestrians, as far as traffic regulations are concerned. Assume moreover that, after a while, pedestrians who become drivers adopt the common inclination of drivers to behave more improperly than pedestrians. Under such conditions, would it make sense to explain the drivers' misconduct by claiming that "driving corrupts"? I take it for granted that every reasonable person will reject such an attempted explanation of the drivers' misconduct. Drivers make decisions with respect to their habits of driving and with respect to particular stages of driving their cars. The roads do not tempt them to misbehave as drivers, just as the pavements do not induce pedestrians to behave properly. Drivers shoulder responsibility for the impropriety of their driving behavior.

Blaming driving as such for drivers' misbehavior is not only vacuous as an attempted explanation of the drivers' behavior. It is also a ludicrous attempt to evade any responsibility for that misbehavior. "It's not me", says the culprit of a traffic accident, "nor my driving decisions, nor my driving style. It is driving as such.". An attempted defense of such a nature would be held absolutely unacceptable.

The same holds true for the attempt to explain military ethical gaps in terms of occupation as such, rather than in terms of habits and decisions of soldiers. The slogan "Occupation corrupts" is a corrupt slogan to the extent that it involves a brazen evasion of responsibility, which is in itself a major ethical gap.

A drunk driver manifests in his dangerous misbehavior something about himself, whether about his values and principles or about his traits of character. There is no reason to assume that such a driver has changed his internal moral constitution when he gets his driving license back. To the same extent, there is no reason to assume that a soldier experiences a shift in his values, principles or character when he assumes the duty of a checkpoint guard. There is a simple explanation for the shifts of behavior when a pedestrian becomes a driver and when a civilian becomes a checkpoint guard. No shift takes place in the deep layers of one's values, principles, and character when one assumes the role of a driver or a checkpoint guard.[9] The basic level governing behavior remains intact and active. What does change is the nature of the situation in which a person is now able to manifest his deeper layers. As a pedestrian, a person did not have an

---

9   It is here taken for granted that commands, ROEs and doctrines never order a soldier to behave at a checkpoint in a way that causes an ethical gap. In case such an unethical (and possibly illegal) command is given, it explains misbehavior and it is in itself an ethical gap.

opportunity to show certain wicked elements of his values, principles, or character, but the moment that person has become a driver, he is in a position to manifest such wicked elements of his deep layers. As a pupil in a high school, a person did not manifest certain elements of his moral nature or personality, but now, at the checkpoint, he does. It is not the checkpoint situation as such which is corrupt; it is rather the humiliating soldier's behavior at the checkpoint which manifests a morally corrupt element of his personal constitution.

At this point, it is sometimes argued that a soldier at a checkpoint should not be blamed for his misbehavior because he has suffered from a weakness of the will.[10] Philosophers have discussed such arguments since the times of Ancient Greece under the title of *akrasia*, which literally means not being in command of oneself. The picture is of a person who "goes into action with all his thoughts intact and available for application but nevertheless acts against them" (Pears, 1982, 1984). To understand the picture of incontinence, think about a deliberation leading to an action (Aristotle had a similar picture in mind in his Nicomachean Ethics). The fundamental principles are the premises of an argument, then there are the different steps of the argument, and finally, as the conclusion of it, we have the action. When the action is incompatible with the fundamental principles, there is some "fault location" inherent in the argument (ibid.). Different explanations of *akrasia* find the fault location at different stages of the argument. Here are some of them.

Notice that the simplest fault location is apparently at some stage of the argument, where an error is made in moving from a premise to a consequence. However, this is a logical weakness, not incontinence. If a soldier acts in a way that is incompatible with his fundamental principles due to logical weakness we face a problem of understanding that can be solved by explanation and controlled experience.

Aristotle explained *akrasia* in terms of emotional interference with reason. A person's appetite for pleasure or anger can make him replace the appropriate conclusion of his argument by an alternative, inappropriate one. One can imagine a soldier at a checkpoint unnecessarily humiliating some person either because he is angry or because there is some pleasure to be gained by acting that way, e.g., manifesting his authority over other persons. Ethical gaps created by an emotional state of anger or by a desire for an emotional state of pleasure involve an ethical gap in ethical education. An adequate ethical education of persons in preparation for a professional activity ought to ingrain in them an effective form of self-restraint. They ought to understand it, form a commitment to manifest it, and then act accordingly.

---

10 Related arguments resort to notions such as "moral fatigue" or "moral erosion".

Another explanation of *akrasia* finds the cause of an action that is incompatible with fundamental principles in a lack of integration. Certain personal desires fail to become properly integrated into the overall system of desires that motivate one's actions. Thus, under usual circumstances actions are performed based on the overall system of desires, but under special circumstances some actions are performed that are independently motivated by a desire which is not part of the overall system of desires.

However, if a soldier at a checkpoint has an independent desire to perform actions that unnecessarily humiliate other persons, then one can rest assured that the ethical gap manifest in the soldier's activity is a clear indication of an ethical gap in the soldier's military ethical education. It is a major goal of military education to shape a system of desires that will motivate a soldier in his or her activities as a soldier. A soldier is entitled to have his own political views, personal attitudes and orientations, private style and taste, but his activity as a soldier is always performed on behalf of the military force, the ruling government and the state, and should therefore reflect orders, ROEs, standard principles of operation, doctrines, policies and professional insights, and not personal elements of the soldier's identity, inclination or mood.[11]

Thus, when we face an instance of misbehavior on the part of a soldier at a checkpoint (and all the more so when we face such instances repeatedly) and the explanation of the weakness of the will is invoked, we may safely assume that if not for an ethical gap at some preceding stage of the soldier's military education, such instances could have never happened, as they should have never happened.

Under a variety of circumstances we have found out that ethical gaps appear in the behavior of soldiers because other ethical gaps took place in their preceding military ethical education. This seems to be a general lesson to be drawn with respect to military education.

Bibliography

Hope, I. (2007). Agility and Endurance: Task Force Orion in Helmand. In K. Patterson & J. Warren (Eds.), *Outside the Wire. The War in Afghanistan in the Words of Its Participants* (pp. 151-163). Toronto: Random House Canada.

Kasher, A. (2002). Between Obedience and Discipline: Between Law and Ethics. *Professional Ethics, 10*(2-4), 97-122.

---

11 For military guidelines that explain this point in the context of the 2006 Disengagement Operation from the Gaza Strip, see Kasher, 2009.

Kasher, A. (2008). The Professional Identity of a Military Force. In A. Mutanen (Ed.), *The Many Faces of Military Studies: A Search for Fundamental Questions* (pp. 17-30). Helsinki: Naval Academy.

Kasher, A. (2009). Military Ethics of Facing Fellow Citizens: IDF Preparations for Disengagement. In D. Carrick, J. Connelly & P. Robinson (Eds.), *Ethics Education for Irregular Warfare* (pp. 87-106). Farnham: Ashgate Publishing.

Kasher, A. & Yadlin, A. (2005). Military Ethics of Fighting Terror: An Israeli Perspective. *Journal of Military Ethics 4*(1), 3-32.

Parks, G., Greenhalgh, T., Griffin, M. & Dent, R. (2008). Effect on Smoking Quit Rate of Telling Patients Their Lung Age: The Step2quit Randomised Controlled Trial. *British Medical Journal, 336*, 598-600.

Pears, D. (1982). How Easy is Akrasia. *Philosophia, 11*(1-2), 33-50. 34.

Pears, D. (1984). *Motivated Irrationality*. Oxford: Clarendon Press.

Stein, J. G. (2003). The Dark Side of Accountability: The Loss of Responsibility. In M. Goldberg (Ed.), *The University Professor Lecture Series 2002-2003* (pp. 15-21). Toronto: Faculty of Arts and Science, University of Toronto.

U.S. Army (1990). *FM 22-100*. Army Publishing Directorate. Retrieved from http://www.enlisted.info/field-manuals/fm-22-100-military-leadership.shtml.

Zhuge, L. & Ji, L. (2002). *Mastering the Art of War* (Thomas Cleary, Trans.). Boston, London: Shambhala.

4. Interoperability and Interculturality

Bo Talerud

Military Leadership for Cooperation? – Education for Civil-Military Cooperation in International Missions

Introduction

International peacekeeping missions have gone through radical changes during the last decades, both in numbers and the nature of the missions. The involvement of civilian actors dealing with political, developmental, humanitarian, human rights, juridical, and social issues has increased rapidly, resulting in a very complex mixture of military and civilian players. There are several obstacles to civil-military cooperation that need to be overcome, for which military leaders at all levels share a responsibility with other actors. In this article I will argue for a more elaborated view of civil-military cooperation and make some suggestions on how to develop a more elaborated approach.

Peacekeeping in international missions: From monitoring ceasefires to supporting the (re)construction of fragile states

Between 1947 and 1989, 15 UN Peacekeeping operations were established. By 2007, 62 UN operations had been launched (United Nations, 2007). In 1998 the UN deployed 14'000 peacekeepers worldwide. In 2007 the number had increased to over 90'000 military and civilian personnel in the field of different UN mandated missions. Also, the nature of these missions has changed fundamentally, from mostly monitoring ceasefires to supporting the reconstruction of fragile states,[1] sometimes even supporting the creation of such states. Moreover, the involvement of civilian actors (both governmental and non-governmental) has increased dramatically. On the whole, there is a complex mixture of military forces from different countries, UN personnel, foreign diplomats, development agencies, other governmental organizations and a great number of various NGOs interacting with local authorities at different levels, local organizations, and the

---

[1] On the concept of 'Fragile states', see Cammac, McLeod, Rocha Menocal & Christiansen (2006).

local population. There are different degrees of disturbances from more or less hostile groups such as rebel factions and their military forces. As Olson and Gregorian (2007) state:

> With the added complexity of such international peace missions and the expanded numbers of international actors and approaches involved, the question of coordination has become a key focus for donors, the UN, other multilateral agencies and NGOs. [...] The trend in peace-building since Kosovo in 1999 is towards greater integration of international efforts and the necessity for collaboration between relief, development and security organizations. (p. 2-3)

However, such integration and coordination is far from easy and dependent on several factors. One is the great number of authorities and organizations involved at different levels. Those different actors have different roles, deployment timelines, procedures, cultures, budgetary pressures, and supervising authorities (United Nations, 2007). In order to just indicate the complexity of this diversity, I will mention a sample of the institutions and processes involved in these international aims to reconstruct fragile states.

The legal basis for the deployment of all UN peacekeeping operations is found in Chapters VI and VII of the UN Charter. After the identification of the need for a UN presence, the analysis of the situation and consultation with the host nation, it is the prerogative of the UN Security Council to determine when and where a UN peacekeeping operation should be deployed. The UN Secretariat in New York is the strategic level in UN peacekeeping and provides policy, guidance and advice to missions (ibid.).

In "integrated missions" between different parts of the UN system, the Special Representative of the Secretary General (SRSG) with his/her office has the overall authority over the peacekeeping activities. The UN system consists, among other actors, of the Department of Political Affairs (DPA), the UN Department of Peacekeeping Operations (DPKO), the Office of the High Commissioner for Human Rights (OHCHR) with its Human Rights Division, the Human Affairs Coordination, the UN Office for Coordination of Humanitarian Aids (OCHA), which coordinates different UN organizations as well as other international organizations, the High Commissioner for Refugees (UNHCR), the UN Force Commander, the Police Commissioner with his/her UN police force, the Information Division, the Food and Agricultural Organisation of the UN (FAO), the World Food Programme (WFP), and the World Health Organisation (WHO).

The institutional structure of the UN system itself is highly fragmented, which is why the coordination between its different parts is not without problems. In the field, the SRSG/Head of Mission (HOM) exercises operational authority over the UN mission's activity in a country (ibid.).

UN missions also have to coordinate and cooperate with external partners, such as bilateral and multilateral donors, international financial institutions, e.g., the World Bank, NGOs or contractors working for donors, the diplomatic corps and other regional and international political actors. Furthermore, cooperation is required with political actors from various nations whose armed forces take part in the mission, and with neutral and independent humanitarian actors, such as the International Committee of the Red Cross (ICRC), Save the Children and a great number of humanitarian NGOs.

These different actors mentioned above have to cooperate with the host nation's national governments, with their regional and local representatives, the host nation's military and police forces and not least, with different parts of the local population. At the same time, the peacekeeping missions sometimes have to deal with the regional and local representatives of rebel factions' political leadership, with the rebel factions' military forces and with populations loyal to the rebels. The missions also have to deal with criminals and other hostile elements, such as warlords and international and local media.

Increasingly, regional organizations like NATO, EU, AU (African Union) and ECOWAS (Economic Community of West African States) are taking part in UN missions (United Nations, 2008). I will mention a few operations where Swedish troops have been involved. Some international operations are led by the UN, e.g., the *UN Missions in Liberia* (UNMIL), or the *UN Interim Forces in Lebanon* (UNIFIL). Other operations are led by NATO with a UN mandate, e.g., the *Kosovo Force* (KFOR) in Kosovo or the *International Security Assistance Force* (ISAF) in Afghanistan. The EU leads a number of operations with a UN mandate, e.g., the *European Union Forces* (EUFOR) in Congo and Chad. These organizations share structures, decision-making and planning processes that are different from the UN system and rather complex in themselves. Participation of those organizations is often vital for the development of the peacekeeping process, but complicates the efforts for cooperation between different civilian actors, different military actors and civil-military cooperation even more.

In the light of all these actors involved in international peacekeeping missions, we can realize the importance of developing the cooperation, coordination, and collaboration of all parts in this complex and disparate mixture involved in peacekeeping operations. Integration is critical for linking the different activities involved in contemporary post-conflict activity (political, developmental, humanitarian, human rights, rule of law, social and security) (United Nations, 2007).

The primary role of UN peacekeeping operations, with regard to the provision of humanitarian assistance, is to provide a secure and stable environment within which humanitarian actors may carry out their activities in a neutral way and independent from political and security agendas (ibid.). In order to achieve the integration mentioned above, there are strong requirements for different competences from the actors involved, including military units. Among other things, competences are required that allow for a cooperation with very different groups of actors, both military and civilian. This cooperation takes place on many different levels: on a central UN, NATO or EU level, on a national level, and on various levels specific to the military. But what do we mean when we say civil-military cooperation or cooperation in general?

Civil-military cooperation and cooperation in general – some definitions

There is no universally accepted definition of civil-military cooperation or coordination (CIMIC) due to a different purpose and focus on the humanitarian and military components. The UN Department of Peacekeeping Operations (UN DPKO) describes UN CIMIC as "the system of interaction, involving exchange of information, negotiation, deconfliction, mutual support, and planning at all levels, between military elements, humanitarian organizations and civilian population to achieve respective objectives" (Lloyd, 2008, p. 53).

The UN Office for the Coordination of Humanitarian Affairs (2003) describes civil military cooperation more from a humanitarian perspective as follows:

> The essential dialogue and interaction between civilian and military actors in humanitarian emergencies that is necessary to protect and promote humanitarian principles, avoid competition, minimize inconsistency, and when appropriate pursue common goals. Basic coordination strategies range from coexistence to cooperation on the spectrum of conflict. Coordination is a shared responsibility facilitated by liaison and common training by the humanitarian and military components. (p. 5)

NATO Civil-Military Co-operation Doctrine (2003) defines the civil-military relationship (CIMIC) more from a military perspective as "the coordination and cooperation, in support of the mission, between the NATO commander and civil actors, including national population and local authorities, as well as international, national and non-governmental organizations and agencies." (p. 1-1).

The NATO doctrine for civil-military cooperation, as it emerged from the 1990s, has been criticized for being rooted in conventional warfare and in a somewhat outdated peacekeeping doctrine (Brocades Zaalberg, 2006). In most current peacekeeping operations with their many and disparate actors, there is a need for every single actor involved to be able to cooperate with different civilian and military actors. For that reason the UN DPKO's definition, together with the OCHA's indication of civil and military *actors* (rather than 'elements' and 'organizations') seems most suitable.

But there are also different degrees of cooperation. As for cooperation in general I will use the following tentative definitions:

- *Cooperation in general:* Common actions with varying degrees of agreement
- *Restricted cooperation:* Exchange and interchange of information
- *Coordination:* of information, intentions, and restricted actions
- *Collaboration:* Formal and informal negotiations, jointly creating rules and relationships, a process toward shared norms and mutually beneficial interaction, where the degrees of agreement might be quite high.

Some short examples of civil-military cooperation models

I would like to give two fairly contrasting illustrative examples of civil-military cooperation. To begin with, Liaison and Monitoring Teams (LMTs) in the Kosovo Force (KFOR) serve as an example of a more restricted cooperation focusing on exchange and interchange of information in a relatively low-conflict mission. In addition, an example of civil-military cooperation with higher ambitions in a more complicated mission is provided by the Provisional Reconstruction Teams (PRTs) in Afghanistan. Both examples are considered with some focus on the participation of the Swedish Armed Forces.

Liaison and Monitoring Teams in Kosovo

NATO's presence in Kosovo rests on a mandate from the UN Security Council (Resolution 1244 from 1999), which stipulates the purpose of the military and civilian international presence in Kosovo in order to "develop democratic institutions", "facilitate the political process", "support the humanitarian organizations" and "protect human rights", all with a view to promoting "financial welfare, stability and regional cooperation". KFOR consists of 5 regional Multi-National Task Forces (Centre, East, North, South, and West) under the Headquarters of the Kosovo Force (HQ KFOR). All together, 34 nations and more than 16,000 peacekeepers are involved. The HQ KFOR reports to the Commander of the Joint Force Command Naples (COM JFCN) in Naples, Italy.

The Multinational Task Force Centre (MNTF C) is deployed in central Kosovo, encompassing Pristina and the villages around. There are KFOR-forces from the Czech Republic, Finland, Ireland, Latvia, Slovakia and Sweden. Sweden has contributed to the KFOR since October 1999, and since that time has been stationed at Camp Victoria in the village of Ajvalija, near Pristina. In 2007, Camp Victoria comprised the following Liaison and Monitoring Teams (LMTs) besides the Swedish contingent: one from the Czech Republic, two from Finland, one from Ireland, one from Slovakia and two from Sweden. In April and September 2007, I visited Camp Victoria, interviewed officers and soldiers in the Swedish contingent and several members of the Swedish LMTs. I also accompanied them during some of their meetings with civilian actors in Pristina and its surroundings.

> Every team consists of five dedicated KFOR-officers whose duty is to monitor the social, political and economical situation in the municipality that they are responsible for. [...] The LMT concept is very straightforward. The officers are highly visible. They have continuous and close contact with the citizens of Kosovo. They work closely with the local administrators; Kosovo Police, UNMIK, community leaders and relevant organizations. They also visit schools, villages and attend a variety of different meetings, both at municipal and regional level (Eriksson, 2005).

In addition to my interviews and visits, I should also mention my contacts with public medical service, local and central politicians, schools and universities, NGOs, international organizations, village leaders, etc. Despite being in uniform, the Swedish LMTs have a rather "civilian" approach in order to optimize relations with these different parts of the civil society. My informants stressed that they do not see themselves as a "CIMIC-organization", as they mostly deal with the exchange of information as a part of military intelligence. All LMTs have the motto "Feel the pulse of Kosovo". They pass the information from the local levels of society on to higher levels in the KFOR, e.g., to the commander of the Multinational Task Force. Naturally, they can also forward information from the KFOR to the local community and even between different actors in the municipality. They do not, however, support the local population with water, roads, computers, etc., but they can inform the CIMIC organizations which villages are in the greatest need for water supply, which schools need computers most, etc.

From my interviews with the LMT members and my visits to different parts of the local communities, I interpret – at least these LMTs – as quite successful and highly appreciated by the local population, but as one of my informants declared: You need to be a special kind of person to be a member of a LMT, i.e., socially competent and highly interested in the culture and customs of the local communities and organizations.

Provincial Reconstruction Teams in Afghanistan

The NATO-ISAF's presence in Afghanistan also rests on a UN mandate and several resolutions from the Security Council.[2] The ISAF's key military tasks include assisting the Afghan government in extending its authority across the country, including conducting stability and security operations in coordination

---

2   Resolutions 1386, 1413, 1510, 1563, 1623, 1707 and 1776 all relate to NATO/ISAF.

with the local authorities. Besides assisting the Afghan government in extending its authority and creating a secure environment, ISAF also supports humanitarian assistance operations.

The political direction and coordination for the mission is provided by NATO's principal decision-making body, the North Atlantic Council. Based on the political guidance from the Council, strategic command and control is exercised by the Supreme Headquarters Allied Powers in Europe (SHAPE) in Mons, Belgium. The ISAF headquarters is based in Kabul and serves as the operational command for the mission. The current five regional commands coordinate all regional civil-military activities conducted by the military elements of the PRTs in their area of responsibility and under operational control of the ISAF. In 2008 there were 26 PRTs operating, five in the Northern region, four in the Western region, four in the Southern and twelve PRTs in the Eastern region. In the Northern region Germany has two, Hungary one, Norway one and Sweden one in Mazar-e-Sharif, containing personnel from Sweden and Finland.

The purpose of the PRTs is to use small civil-military teams to expand the legitimacy of the central government to the regions and enhance security by supporting the security sector reform[3] and facilitating the reconstruction process of society. This implies a mixture of cooperation levels (i.e., restricted and more elaborated cooperation). They consist mostly of military personnel (90-95% of total), political advisors and development experts.

The PRTs are lightly armed for self-defense only and remain more of a diplomatic than a military tool, being a "robust military diplomacy". PRTs are not there to conduct humanitarian operations of their own but to facilitate the activities carried out by international organizations and NGOs (Jakobsen, 2005).

There are at least three different PRT models: a US, a UK, and a German model. The US PRT operates under military command, adopting a robust approach, strongly emphasizing force protection. The UK PRT consists of a joint leadership between the military, political, and developmental components. The British military has been less directly involved in reconstruction efforts compared to the US PRT. The civilian components run their programme with minimal involvement from the military and report back to their respective organizations (ibid.).

> UK PRTs have enabled the British to establish a better relationship with NGOs than the Americans. [...] Compared to the American and British models, the German model includes a higher number of civilian personnel and a higher degree of separation as the civilian personnel are not under military command. (p. 23-24)

---

3 According to Stapleton (2007), the Security Sector Reform focuses on disarmament, demobilisation and reintegration (DDR), the creation of a new Afghan army, a reformed police force, judicial reform and counter narcotics strategies.

The military focus is strong in the US PRTs while the British model has a joint civil-military leadership and the civil part has more autonomy than the US model. The German model is characterized by civil-military separation.

According to Jakobsen (ibid.) the British-led PRT has been more successful than the German and the US-led PRTs:

> The UK formula for success has, in short, consisted of extensive consultation and cooperation with all the relevant actors in the area, a willingness to heed NGO and UN advice, a strong focus on security, effective intergovernmental cooperation, in-depth understanding of local security dynamics, and a robust approach towards spoilers coupled with extensive long-range soft patrolling aimed at winning hearts and minds. (p. 33)

Maybe, this British PRT model has some resemblance with the Canadian concept of "3D". This concept requires that the various agents of government (Development, Diplomacy and Defense) work together to achieve a set of common all-embracing goals concerning security and development, for which the Asian tsunami relief in Sri Lanka provided the best 'test case'. I think that the idea of this concept is to develop collaborations between the 3 Ds on equal terms, without one dominating the other. Maybe the Canadians will also try to adopt this model for their PRT.

Sweden, among other countries, has adopted the British model. The following guidelines were issued to direct the Swedish-Finnish PRT in Mazar-e-Sharif (Sanberg, 2008):

- Civilian success is our most important military aim
- Achieve credibility by presence
- Be prepared for the hostile few, but act to strengthen the consent of the others
- Get effect through the UN, the GoA (Government of Afghanistan), the Local Authorities, and the NGOs
- Know the key player in our AOR (Area of responsibility)
- Do respect but do not imitate the Afghan culture and religion
- Listen, be patient and humble; we are guests
- Know common phrases of courtesy
- Respect and encourage human rights

It seems that one of the most important activities for the Swedish-Finnish PRT is patrolling far out in the area of responsibility with "MOTs" (Military Observation Teams) consisting of two officers, four to six soldiers with different specialties (e.g., radio operator, driver, nursing attendant, etc.) and an interpreter. They speak to different local leaders, e.g., political, religious or police officers, teachers, etc., in order to get an idea of the situation regarding security and development in the area. As the MOTs are both the "ears and eyes" and the spokesmen of and for the ISAF, they resemble, to some degree, the LMTs in Kosovo. But there are of course also considerable differences. The ambition of the PRT model to expand the legitimacy of the central Afghan government, to support a sector reform, and to facilitate the reconstruction process of an even more fragile society than Kosovo renders the PRT's work not only more dangerous but also much more complicated and therefore more vulnerable for criticism.

Generally speaking, the PRT model is seen as the key mechanism for security and reconstruction in Afghanistan outside Kabul, and has received a lot of attention as a means of coordinating civilian and military efforts in the reconstruction of the Afghan society. But the PRT concept has also faced some criticism, from being successful though not sufficient, to merely obstructing efforts of reconstruction and rebuilding. There are critics who claim that the PRTs have not produced the intended development in either security or reconstruction.

In March 2007, there was a major workshop for researchers and practitioners at the University of Calgary's Centre for Military and Strategic Studies. Olson and Gregorian (2007) elaborate:

> Thirty-five representatives of key civilian and military assistance actors involved in Liberia and Afghanistan together with experts on peacebuilding examined whether current coordination efforts amongst diverse assistant actors effectively support the interlinked goals of supporting security, development and sustainable peace. (p. 19)

The lessons learned in Afghanistan, according to the workshop, resulted from tensions between military and development actors over a perceived "militarization of aid" and the merging of military and assistance agendas within the PRTs (ibid.).

> Examples of effective coordination cited were the health sector and the government's National Solidarity Programme, where participatory, inclusive processes involving the government, donors and implementing NGOs has resulted in significant 'buy-in' from all key stakeholders. The record of coordination in many security-related areas however is poor, as evidenced by a lack of coordination on counter-narcotics and diverse national strategies for the PRTs. [...] Furthermore, there were strong disagreements over appropriate *means* to conduct programmes, and *few common principles* guiding efforts of military and development actors. More fundamentally, current coordination processes ignore differences in power and influence. [...] In the field, the perception is that the perspectives of military forces dominate and dramatically affect the work of aid organizations. (p. 19)

By virtue of its resources and political weight, US policy often leads behind the scenes. That becomes a complicating factor for many development actors coupled with a suspicion that coordination efforts on offer could be simply an attempt by powerful players to exert control over the activities of smaller actors (ibid.).

Obstacles to civil-military cooperation

As indicated in the previous section, there are lots of obstacles, not only to the PRTs, but also to many forms of civil-military cooperation[4]:

- Lack of common purpose: Different actors can disagree on the purpose and the degree of effort in the reconstruction and cooperation. According to the Canadian workshop mentioned above, common aims regarding civil-military coordination in the field were largely absent.
- There are tensions between different agents of government due to different cultures, agendas, interests, routines and guiding principles.
- There are tensions between ministries, public authorities, and the field due to different routines, culture, and opinions about adequately fast feedback.
- Some NGOs have difficulties or even refuse cooperation with military units, as they think that their security rather decreases than increases when they become mixed up with military forces.
- The humanitarian aid sector, with which the military must operate on the ground, often has just a superficial understanding of the military perspective.
- The military presence being the strongest fraction results in its perspectives, culture (formed on the traditional "just war concept"), and perception of the cooperation as a predominantly military operation to dominate. Combat perspectives and traditional military identity dominate over developmental perspectives in many military units.
- There is a different sensitivity to power asymmetries inherent in coordination efforts.

What are the demands on military leaders?

Taking those obstacles into consideration, it is of course of great interest how we can develop civil-military cooperation in connection with peacekeeping operations. One part of it consists of the art of leadership. In their Peacekeeping Operations Principles and Guidelines, the UN (2007) emphasize the importance of good leadership in peacekeeping operations:

> The selection of the mission leadership must be a carefully considered process. While doctrine and guidelines may provide the conceptual framework for an integrated mission, it is the moral strength, example and guidance of the senior leadership which will cement the components together and develop a unity of effort. (p. 43)

---

4 The list of obstacles was put together with the help of the following sources: Cooter, 2007; Fors & Larsson, 2007; Okros, 2007; Olson & Gregorian, 2007.

Discussing "spheres for improvement" in the civil-military cooperation process, Olson and Gregorian (2007) stress that "properly vetted and trained personnel, who can exercise the kind of leadership that cross-organizational and cross-cultural collaboration requires, be placed in leadership positions at all levels" (p. 32-33).

I think it is important that not only the senior leadership, but all individuals "in leadership positions" and even their collaborators are "properly vetted and trained" for cross-organizational and cross-cultural complex civil-military cooperation in connection with peacekeeping operations and the reconstruction of fragile states. Out in a PRT MOT mission even a single soldier can encounter situations where he or she has to act as a "leader" in interactions with the local population or fellow soldiers.

Generally speaking, there are two main approaches to defining leadership, the first one being trait definitions of leadership, where leaders are defined based on intelligence, extroversion, fluency, charisma, etc. Trait-defined approaches are often characterized by top-down communication. The other approach consists in process definitions of leadership that emphasize the interaction between leaders and collaborators (Northouse, 2007). Some leadership research emphasizes the reciprocity in this interaction even more (Baker, 2007; Pearce & Conger, 2003). Even if traits sometimes do have some influence I will emphasize process definitions of leadership such as the following (Yukl, 2008): "Leadership is the process of influencing others to understand and agree about what needs to be done and how to do it, and the process of facilitating individual and collective efforts to accomplish shared objectives" (p. 8).

So how can we facilitate individual and collective efforts to develop civil-military cooperation in spite of the many obstacles mentioned above?

How to counter some of the obstacles to civil-military cooperation

First, I think it is important for everyone involved in a peacekeeping operation to focus on the overall purpose, which is to provide a secure and stable environment according to existing UN mandates in order for humanitarian and development actors to carry out their activities. Some efforts should be made to discuss the purpose and degree of the societal reconstruction work and cooperation with different actors. In dealing with actors in the host nation, both soldiers and officers would still be "firm, fair, and friendly", but also devoted to human security and the process of facilitating the reconstruction of a fragile society by civil actors.

Elron (2008) explains the various power asymmetries in a peacekeeping operation as "an unequal distribution of power, with the more powerful actors dominating the council, the UN and the world's political system" (p. 32). Moreover, there is also an unequal distribution of power and influence among western and non-western military units, among native English speaking units and non-English speaking ones, etc. (ibid.). Most notable for NGOs is what they experience as a "militarization" of missions, due to a disproportion in the number of personnel, but also due to the strong military culture and its impact on traditional "combat skills identity". Communication with NGOs would probably be facilitated if the respective awareness increased among members of the armed forces, including some respect and understanding for different agendas, ideology, and cultures inherent in many NGOs. Traditional military identity and combat skills should be transformed into a new kind of military identity and self-understanding in which the necessary combat skills are combined with contact skills. As Nørgaard and Holsting (2006) state, "contact skills and combat skills should never be perceived as mutually *exclusive*, but rather as mutually *reinforcing* competencies" (p. 131).

In order to be able to show *the capability* of combat skills in a convincing way, it is probably important to gain respect and confidence from different actors, which is many times the prerequisite for developing contact with parts of the host nation's population. But the more contact skills you have, the more mutual trust you can develop, and the less often you have to actually use your combat skills. For that reason, both of these mutually reinforcing competencies are equally important qualities in a peacekeeping soldier's identity. The development of contact skills, including the understanding of different agendas, ideology, and cultures, is naturally also important in contacts with other military units and different kinds of civilian actors – among them the different NGOs.

In addition to this more elaborated military identity, soldiers and officers may perhaps have an educational task when arguing in favor of military perspectives and reasons and explaining them to those who only have a superficial understanding of them.

Steps to a more elaborated capacity for civil-military cooperation – some suggestions

Efrat Elron, a research fellow from the Hebrew University in Jerusalem, has suggested some "intercultural integrating mechanisms" for the interplay between military units in multinational UN peace operations (2008). Some of them could perhaps apply to civil-military cooperation as well. Among the suggested mechanisms are:

- *Knowledge sharing and mutual learning.* Civilian and military actors involved in peacekeeping operations all have different experiences and knowledge. As a consequence, susceptibility for mutual learning from different perspectives, experiences, and cultures could stimulate creative thinking concerning one's own perspective and preconceptions and create opportunities for innovations in actions, procedures, and organization.
- *Joint training* between civilian and military actors before and during missions could increase mutual understanding.
- *Cohesion building activities* in the form of common educational, cultural, and sports events, but also drinking and/or eating together could increase mutual trust and thus create opportunities for cooperation, affiliation, and mutual informal learning under relaxed circumstances.

Capabilities to develop constructive ways for interaction with a great variety of civil actors (local leaders, NGOs, UN personnel, etc.), and competence to organize constructive meetings between military and non-military actors will of course make the security and reconstruction process easier. Taking into account the complex and disparate variety of actors and their different agendas, cultures, supervising authorities, etc., it all boils down to a need for soldiers and officers to develop a *multidimensional cultural awareness and competence* enabling them to interact constructively and better understand the vast array of actors. This includes not only different kinds of contact with the local population in the host country, but also contact with UN personnel, political advisors, development experts, actors from different NGOs, the differences among the involved nations, etc. (Northouse, 2007; Selmeski, 2007; Yukl, 2008).

In order to develop such multidimensional cultural awareness and competence, a part of basic education and training for soldiers and officers should be to handle different kinds of civil-military cooperation, including basic cultural theory from social anthropology respectively ethnology and multidimensional cultural studies and training. The focus would be on learning *how* to think about cultural matters rather than *what* to think, thus allowing asking questions and perceiving outside the box. "The complex nature of culture will require increased emphasis on education rather than the military's traditional reliance on experience, training, and self-development" (Selmeski, 2007, p. 21). Within such basic education, there should also be opportunities to develop an elaborated military identity, including a cultural self-identity that focuses not only on combat skills, but also on contact and developmental skills and competence to act as a cross-cultural facilitator. As an additional element of this basic formal military education, it could be useful to have some courses and training with civilian actors engaged in developmental work in fragile societies. Of course this is just a tentative idea that needs to be developed much further.[5]

In the Canadian Forces it is widely recognized that cultural awareness is of paramount importance in different peacekeeping operations and particularly in Counter Insurgency Operations. The Department of National Defence in Canada has thus formalized a partnership agreement with the Canadian Centre for Intercultural Learning (CIL) implementing training for all military personnel deploying to Afghanistan and other international missions (Séguin & Savard, 2008).

> Since implementing this training approach in January 2008, over 1'600 military personnel have participated in cultural awareness training delivered by CIL. Overall feedback and comments indicate that this blended learning approach, including the use of critical incidents, relevant subject matter expertise and e-learning resources are best suited to meet the needs of this diverse audience. (p. 30)

Moreover, defense forces in other countries can perhaps establish cooperation with civilian intercultural or multicultural centers for multidimensional cultural training as well.

In connection with multidimensional studies and training, one should also pay attention to the differences between *teamthink* where actors contribute with their different experiences and knowledge in order to accomplish shared objectives, and the danger of *groupthink*, which is characterized by mental introversion, a pressure against dissidents, and overestimating one's in-group and its perspectives, power, and morals (Granberg, 2004, 2006).

Before each mission there should be an increased emphasis on the UN mandate and the overall purpose of the mission, the importance of multidimensional cultural and ethical awareness and on human security (Talerud, 2007).

Challenges for continuous developmental learning

As important and necessary formal pre-mission education and training are, they merely provide basic skills and a conceptual ground for insights and understanding. No less important is continuous non-formal and informal learning during the mission, as Nørgaard and Holsting (2006) point out:

> A recurrent theme of these personal military narratives deals with showing the right 'attitude' or, more specifically, a 'professional attitude' expressed not only through professional

---

[5] E.g., Selmeski (ibid.) outlines suggestions on how to enhance cross-cultural competence within the Canadian Military Professional Development system, ranging from the enhancement of the cultural self-identity of citizen-soldiers to higher officers' ability to act as cross-cultural ambassadors.

expertise and basic soldiering skills but also through the will to *improve oneself* continually. 'Professionalism' thus embraces the full breadth of an individual's character and moral outlook – i.e. the values he represents and exemplifies in the way he presents himself. In this context, 'professionalism' is a military formative ideal [...]. (p. 65)

The emphasis on the soldier's character and moral outlook seems to have some similarities to the concept of "Action Competence" as a "psychological-pedagogical-ethical-political term tightly connected with the idea of democracy" and the Aristotelian concept of *phronesis* as practical wisdom (Toiskallio, 2004). Applied to military professionalism as a "formative ideal" it underlines a broadening and elaboration of the competencies in the military profession, stressing the need for continuous learning and development.

During the mission, an enquiring, critical, and autodidactic systematic reflection on psychologically challenging experiences as well as on the benefits of mutual learning from different perspectives, experiences, and cultures is of vital importance. According to the Swedish professor Per-Erik Ellström (2002, 2004), learning processes within organizations could be of at least two different kinds, connected to two logics of activity and learning:

- *Production-oriented learning*, in accordance with the production of logic, and with an emphasis on practiced and effective action, problem solving through application of given rules and instructions, consensus, standardization, stability and avoidance of uncertainty. Adaptive learning oriented towards the mastering of procedures and routines.
- *Developmental learning*, in accordance with the logic of development, and with emphasis on thought and reflection, alternative thinking, experimentation and risk taking, tolerance for ambiguity, variation and mistakes; development-oriented learning.[6]

I suggest that the complementary use of these two modes of learning is of paramount importance for international missions and civil-military cooperation. Both types are of equal importance and the challenge is to develop the most fruitful combination of the two. It is a great challenge for military leaders to facilitate a constructive combination and dynamics between production-oriented and development-oriented learning in such a way that it suits both the leaders and their co-workers. With Ellström's two types of logic in mind, we can thus distinguish two types of (complementary) leadership: on the one hand there is *production-oriented leadership*, where the leader acts as a "coach", inspiring the co-workers to perform better and more efficiently in favor of the ongoing "production". On

---

6   There are similarities between production/development-oriented learning and the distinction between single-loop- and double-loop-learning made by Argyris & Schön, 1978.

the other hand, there is *development-oriented leadership,* where the leader inspires reflection, alternative/creative thinking and developmental learning. As both types are necessary, there can be a constructive mixture of these leadership modes in the same person.

Following this line of argument, I suggest that leaders have special obligations to encourage self-education among subordinates and should therefore have some insights into the practice and theories of non-formal and informal "on-the-job" learning (see Candy, 1991).[7] Efforts should be made so that experiences from the field might influence formal military education and organizational learning within the military system.

The military organization should develop strategies to increase the knowledge and support of organizational learning and re-learning (Argyris & Schön, 1978; Cohen & Sproull, 1996; Elkjaer, 2003), so that experiences from missions of civil-military cooperation – with their possibilities and obstacles – can influence the military system and its formal education, contributing to a more systematic developmental approach to civil-military cooperation. At least higher officers should also have some motivation to support organizational learning and re-learning within the military system towards an elaborated identity within the military profession. Taking into account the downsizing of many armed forces, education of higher officers in organizational learning and re-learning could be arranged at regional levels, e.g., Nordic, European, NATO, etc.

In my opinion, civil-military cooperation, along with the intentions of the international society to support the (re)construction of fragile and vulnerable states, has become extremely important. However, it is also difficult and complicated to accomplish, as I hope this article has indicated. There are of course political, economical, ideological, cultural, etc., reasons for this, but military organizations can contribute not only at field levels but also along the organizational hierarchy – provided they are appropriately educated and trained.

Bibliography

Argyris, C. & Schön, D. (1978). *Organisational Learning.* Reading: Addison Wesley.

Baker, S. (2007). Followership: The Theoretical Foundation of Contemporary Construct. *Journal of Leadership & Organisational Studies, 14,* 50-60.

---

[7] During the last decade there have been many articles, pamphlets, books and research on "lifelong learning" and workplace related learning. See also Ellström & Hultman, 2004.

Brocades Zaalberg, T. W. (2006). Countering Insurgent-Terrorism: Why NATO Chose the Wrong Historical Foundation for CIMIC. *Small Wars & Insurgencies, 17*(4), 399-420.

Cammac, D., McLeod, D., Rocha Menocal A. & Christiansen K. (2006). *Donors and the 'Fragile States' Agenda. A Survey of Current Thinking and Practice* (Report submitted to the Japan International Cooperation Agency). Poverty and Public Policy Group, Overseas Development Institute, London, UK.

Candy, P. (1991). *Self-direction for lifelong learning.* San Francisco: Jossey-Bass.

Cohen, M. & Sproull, L. (Eds.) (1996). *Organizational Learning.* London: SAGE.

Cooter, C. (2007). A Canadian perspective on civilian-military cooperation: Making a good idea better. In *Civil-Military Cooperation in Multinational Missions Conference Report* (p. 18-25). Stockholm: Forum for Security Studies, Swedish National Defence College.

Elkjaer, B. (2003). Organisational learning with a pragmatic slant. *International Journal of Lifelong Education, 22*(5), 481-494.

Ellström, P.-E. (2002). Time and Logics of Learning. *Lifelong Learning in Europe, 2,* 86-93.

Ellström, P.-E. (2004). Reproduktivt och utvecklingsinriktat lärande i arbetslivet. In P.-E. Ellström & G. Hultman (Eds.), *Lärande och förändring i organisationer – Om pedagogik i arbetslivet.* Lund: Studentlitteratur.

Elron, E. (2008). The interplay between the transnational and multinational – Intercultural integrating mechanisms in UN peace operations. In J. Soeters & P. Manigart (Eds.), *Military Cooperation in Multinational Peace Operations* (p. 28-48). London, New York: Routledge.

Eriksson, S. (2005, January 31). Closer cooperation with new concept. MNB (C) introduces Liaison and Monitoring Teams. *KFOR Chronicle.* Retrieved from http://www.nato.int/kfor/chronicle/2005/chronicle_01/03.htm.

Fors, M. & Larsson, G. (2007). *Civil-militär samverkan på departementsnivå – påverkan på fältnivån i Afghanistan [Civil-military cooperation on government ministry level – influences on field level in Afghanistan].* Stockholm: Swedish National Defence College.

Granberg, O. (2004). *Lära eller läras – Om kompetens och utbildningsplanering I arbetslivet.* Lund: Studentlitteratur.

Granberg, O. (2006). Arbetslag och kollektivt lärande. In M. Lindholm (Ed.), *Pedagogiska Grunder* (pp. 169-197). Stockholm: Försvarsmakten.

Jakobsen, P. (2005). *PRTs in Afghanistan: Successful but not sufficient* (DIIS REPORT 2005: 6). Copenhagen: Danish Institute for International Studies.

Lloyd, G. (2008). *An exploratory study for the psychological profile of a civil military coordination officer as a selection tool for training.* Unpubl. MComm thesis, Dept. of Industrial Psychology, University of Stellenbosch.

NATO (2003). *Allied Joint Publication 9.* Retrieved from http://www.nato.int/ims/docu/AJP-9.pdf.

Nørgaard,K. & Holsting, V. (2006). *International Operations in FOKUS.* Copenhagen: Royal Danish Defence College.

Northouse, P. (2007). *Leadership – Theory and Practice.* Thousand Oaks: SAGE.

Okros, A. (2007). 3D Security: The Implications of integrating defense, diplomacy and development in multi-national missions. In W. Patoka (Ed.), *Civil-Military Cooperation in Multinational Missions Conference Report* (pp. 7-17). Stockholm: Forum for Security Studies, Swedish National Defence College.

Olson, L. & Gregorian, H. (2007). Interagency and Civil-Military Coordination: Lessons from a Survey of Afghanistan and Liberia. *Journal of Military and Strategic Studies, 10*(1).

Pearce, C. & Conger, J. (Eds.) (2003). *Shared Leadership – Reframing the Hows and Whys of Leadership.* Thousand Oaks: SAGE.

Sanberg, G. (2008). *Cooperation with NGOs. Guidelines for the Swedish/Finnish ISAF-force.* Presentation held at the Swedish National Defence College on April 22nd, 2008, Stockholm.

Séguin, R. & Savard, D. (Eds.) (2008). New Collaboration between DND and the Centre for Intercultural Learning. *Intercultures Magazine, 4*(2), 29-30.

Selmeski, B. (2007). *Military Cross-Cultural Competence: core concepts and individual development* (Contract Report 2007-01). Air Force Culture and Language Center.

Stapleton, B. (2007). A means to what end? Why PRTs are peripheral to the bigger political challenges in Afghanistan. *Journal of Military and Strategic Studies, 10*(1).

Talerud, B. (2007). Ethos and Ethics for Today's Armed Forces in Western Societies. In J. Toiskallio (Ed.), *Ethical Education in the Military: What, How and Why in the 21st Century. ACIE Publications* (p. 63-84). Helsinki: National Defence University.

Toiskallio, J. (2004). Action Competence Approach to the Transforming Soldiership. In J. Toiskallio (Ed.), *Identity, Ethics, and Soldiership. ACIE Publications* (pp. 107-130). Helsinki: National Defence College.

UN Office for the Coordination of Humanitarian Affairs (2003). *Guidelines on the Use of Military and Civil Defence Assets to Support United Nations Humanitarian Activities in Complex Emergencies.* Geneva: OCHA.

UN Security Council (1999). *Security Council Resolution 1244 (1999) [on the deployment of international civil and security presences in Kosovo]*. Retrieved from http://www.unhcr.org/refworld/docid/3b00f27216.html.

United Nations (2007). *UN Peacekeeping Operation. Principles and Guidelines (Capstone Doctrine Draft 3 29 June 2007)*. UN Dept. of Peacekeeping Operations, Dept. of Field Support.

United Nations (2008). *Guidelines for joint UN-EU planning applicable to existing UN field missions*. UN Dept. of Peacekeeping Operations, Dept. of Field Support.

Yukl, G. (2006). *Leadership in Organizations*. New York: Pearson Prentice Hall.

Juha Mäkinen

Constructively Aligned Military Education and Training in Finland in the Times of the European Bologna Process

The main premises of this paper are that the *aims* of academic degrees are set both internationally and nationally through the evolving Bologna Process, allowing each military educational organization to apply them in an *interactional* way to their practices. Secondly, the aims of the degrees and the specified learning outcomes of the studies are *not reducible to each other*. In the times of the Bologna Process it has been widely claimed that we should consider defining the (core) competencies, skills, and abilities to be taught and consequently learnt (i.e., internalized) at the different levels of the military hierarchy. But what then, if any, is the meaning of the aims of the degrees? This paper concentrates on the practical meaning of the aims, while attention is paid to the shared aims and objectives, principles, and visions of higher education in Europe, but also on another level, to the intended learning outcomes.

When the aims of the degrees, as well as the intended "learning outcomes", are defined by civilian authorities, the consequent key question for military educational organizations is the military pedagogical meaning of the Bologna principles, models, and conceptualizations (e.g., descriptors). In this paper it is emphasized that the Bologna Process does not enter a "local vacuum", as each local context has its specific culture, apart from an educational and a pedagogical culture, influencing how the educational transformative processes will develop. This means that in principle, and in practice, as the Finnish case analyzed in the paper shows, the Bologna Process has been translated, interpreted, and acted upon in various ways. The present paper is not only a historical analysis of the Bologna Process in Europe, in Finland, and FNDC/FNDU[1], but its main intent is to be a future-oriented analysis when the (military) pedagogical meaning of the Process keeps being renegotiated among military institutions of higher learning.

---

1 The abbreviation FNDC stands for the Finnish National Defence College and FNDU for the Finnish National Defence University. On January 1st, 2007, the FNDC was renamed in English as the Finnish National Defence *University*.

A brief overview of evolution of the European Bologna Process

The so called Bologna Process has been going on in Europe for over 10 years (Benelux Bologna Secretariat, 2009; Mäkinen, 2005). As many of us remember, the Bologna Process is named after the joint Bologna Declaration, which was signed in the Italian city of Bologna in 1999 by ministers in charge of higher education. In that declaration *the main objectives* were set for the process of establishing and promoting the European area of higher education (European Ministers of Education, 1999). The meeting in Bologna was followed by many ministerial meetings in several European cities, and each time an official communiqué was written. The analysis presented here focuses on the military pedagogical meaning of them, at least to some extent.

While keeping the main Bologna principles in mind, we should analyze the Prague Communiqué (European Ministers of Education, 2001) as well. In Prague the ministers, with their supporting staff, promoted European cooperation in quality assurance. The ministers recognized the vital role that *quality assurance systems* play in ensuring high quality standards and in facilitating the comparability of qualifications throughout Europe. The ministers called upon universities and other higher education institutions, national agencies, and the European Network of Quality Assurance in Higher Education (ENQA), in cooperation with corresponding bodies from countries that are not members of ENQA, to collaborate in establishing *a common framework* of reference and to disseminate best practices (European Ministers of Education, 2001).

In Berlin in 2003, the ministers encouraged the member states to elaborate a framework of comparable and compatible qualifications for their higher education systems, which should seek to describe qualifications in terms of workload, level, learning outcomes, competences, and profile. They should also attempt to elaborate *an overarching framework* of qualifications for the European Higher Education Area (EHEA). Within such frameworks, degrees should have different *defined outcomes*. First and second cycle degrees should have different orientations and various profiles in order to accommodate a diversity of individual, academic, and labor market needs, as stated in the Berlin Communiqué (European Ministers of Education, 2003).

The history of the Tuning project began already in 2000 (Gonzales & Wagenaar, 2008). It was reflected in the Berlin Communiqué as serving as a common basis (i.e., shared language) for an overarching European framework of qualifications. It has often been overlooked (Kallioinen, 2008; Paile, 2008) that the Tuning project is not only about a shared competence language, but more about shared visions, methodologies, methods, and models for European educational organizations. As Figure 1 shows, the Tuning project aims to elaborate approaches to learning, teaching, and assessment as part of the functioning of the quality enhancement in the educational processes. So does educational and pedagogical research, and in local institutions these two approaches interact with each other, but in what ways?

When rereading the material of the Tuning project, and the several Communiqués before and after the ministerial meeting in Berlin, one recognizes that the Tuning project has meant not only that specified learning outcomes should be described for the written curricula, but also that the local activities of teachers and students have a pivotal role in the process of transforming European higher education. Consequently, a more holistic view of the curricula is needed, meaning that also the taught, and learned, curricula need to be taken into account (Mäkinen, 2006).

Figure 1. The Tuning Model

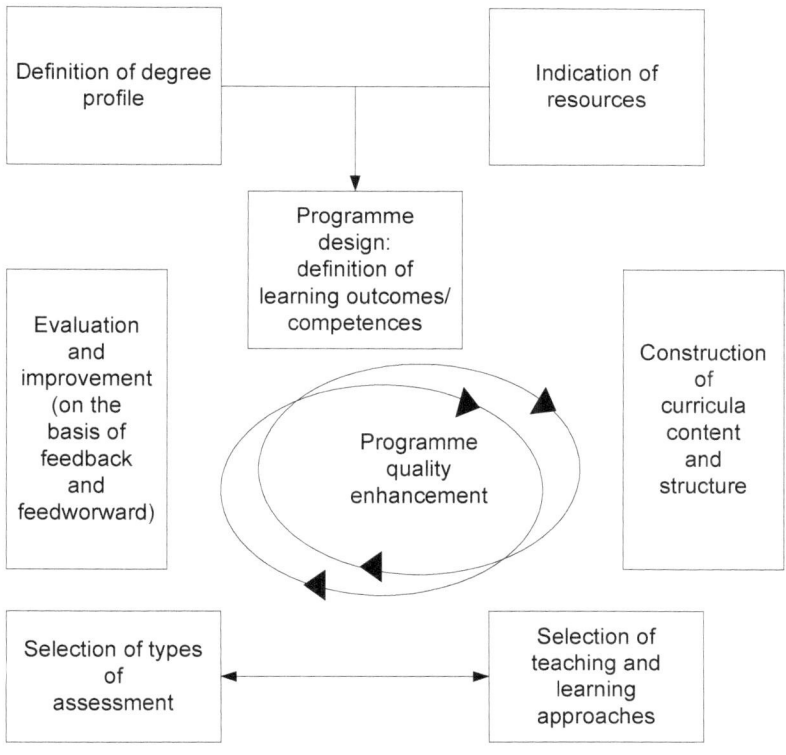

(Gonzales & Wagenaar, 2008)

In 2005, the ministers of education agreed in Bergen that they will adopt the overarching framework for qualifications in the EHEA, comprising three cycles (including, within national contexts, the possibility of intermediate qualifications), *generic descriptors* for each cycle based on learning outcomes and competences, and credit ranges in the first and second cycles. They committed themselves to elaborating national frameworks for qualifications compatible with the overarching framework for qualifications in the EHEA by 2010 (European Ministers of Education, 2003).

The first set of so called Dublin descriptors for both lower and higher university degrees were first proposed in 2002 by an informal network for quality assurance and accreditation, namely the Joint Quality Initiative (Joint Quality Initiative, 2009). The completed set of descriptors was proposed in 2004, differentiating the three cycles of higher learning in terms of specific criteria. Again, an alternative seed for a Europe-wide competence language was planted. But as stated above, we cannot reduce the pedagogical meaning of the Bologna Process to the level of shared competence language only; instead the meaning of the Process is much more holistic (e.g., military pedagogical). Before continuing this brief narrative of the Bologna Process, and especially of its military pedagogical meaning for European educational organizations, I will turn to the specific case of Finland.

Identifying the military pedagogical meaning of the Bologna principles in Finland

The implementation of higher education reforms takes place in certain times and places. The context of the implementation is important if we wish to understand the reform in a national system of higher education (Hoffman, Välimaa & Huusko 2008; Tomusk, 2007). There are currently 17 universities and 27 universities of applied sciences in Finland: one higher education institution per 123'000 inhabitants. This ratio indicates that education, especially higher education, is highly appreciated in Finland, and the expansion of this system in the 1960s through the 1990s was closely linked with a strong welfare-state agenda (Välimaa, 2001).

In the initial phase of the Bologna Process (1999-2000), the Finnish Ministry of Education had to "sell" the idea by focusing on general and European problems in higher education[2]. The Bologna process was presented as an answer to these

---

2   About the initial problems to be solved with the Bologna Process, see Lehikoinen (2001) referred to by Välimaa, Hoffmann & Huusko (2007, p. 46-47). The crucial question is whether these "civilian problems" are the same as experienced and interpreted in the military educational organizations? Is it central for military organizations to "sell" anything, or just to execute the orders given by superiors? Of course the military organizations have to "sell" many things, not

problems. This initial step was necessary, as Finnish academics were skeptical of the Bologna Process. The follow-up oriented studies of the Bologna-process (e.g., Haug & Tauch, 1999; Hoffman, Välimaa & Huusko, 2008; Lourtie, 2001; Mäkinen, 2006) confirm the conclusion that the Bologna process had been translated and acted upon by local actors and these translations can be imagined along a spectrum from an overall acceptance of the Bologna process to resistance towards it[3]. The conclusion should not surprise us since each of our organizations is rooted in their own cultures which need to be taken into account when the educational processes are planned and implemented.

When the Bologna process was initiated and about to cover the field of European educational institutions, the basic training for warrant officers was discontinued in 2001 in Finland and a two-tier degree structure[4] was adopted for the officer education programme in the Finnish Defence Forces (FDF) (Government of Finland, 2008). The educational system of Finland continued its evolution, and due to the reform process of the Finnish polytechnic sector (Ministry of Education, 2002), the role of the former Warrant Officer School (*Päällystöopisto* in Finnish) changed, and the Army Academy was established as a part of the Finnish officer education system (Finnish Defence Staff, 2002).

In 2004, the new Decree on University Degrees was issued (Finnish Ministry of Education, 2004), and it set the general aims for the Bachelor and Master level degrees to be applied in the FNDC/FNDU as well. A brief analysis of the military pedagogical meaning of the aims for university studies is presented below. In Finland, the studies leading to either a lower (i.e., the first cycle degree; BA) or a higher (i.e., the second cycle degree; MA) degree will provide the students with, e.g., knowledge of the fundamentals of the major and minor subjects or corresponding study entities or studies included in the degree programme, and the prerequisites for following the development in the field. For all the students at the FNDC/FNDU this has meant and means that he/she has to choose a major and minor academic subject. The students have to choose whether their area of expertise is the art of war (containing its own basic fields of operational art and

---

    only to their personnel, but also to society in general, and these military academic issues are not an exception to this principle.

3   Hoffman, Välimaa & Huusko (2008) offer an explanation why individual academics in units on the same campus perceive the same set of reforms in such different ways. Their answer emphasizes the centrality of the different *competitive horizons*, i.e., interpretations of their most serious competitor ranging from global (type 1) to micro/national (type 3), when type 2 orientation is somewhere in the middle (i.e., potentially global orientation).

4   The officer's lower academic degree was equivalent to 120 credit units (i.e., study weeks) and consisted of about three years, and the officer's higher academic degree was equivalent to 160 credit units, consisting of about four years.

tactics, strategy, and military history), military pedagogy, leadership, or military technology. Additionally, it has been ordered that each student has to choose the art of war either as his/her major or minor academic subject. After completing their BA studies, the students are allowed to change their major and minor subjects, and consequently, in some academic subjects, additional studies are needed before the beginning of the MA studies.

Both the lower and higher university degrees offer the student knowledge and skills needed for scientific thinking and the use of scientific methods. Therefore the core of the studies consists of the students' research processes mainly in the form of a BA or MA thesis. Also many other studies in the BA, and especially in the MA courses, are done by researching progressively. The centrality of this principle is even more emphasized when it is noted that the university degrees provide knowledge and skills not only for studies leading to postgraduate education, but also for operating independently as an expert and developer of the field. It follows that the fundamental question about the meaning of *expertise* is also a military pedagogically interesting one. Finally, all students of the BA and MA courses should have adequate language and communication skills in Finnish, Swedish (Finland has two official languages) and one other language; most of the students study English, and in military pedagogical MA studies some of the courses are taught in English (cf. Decree of University Degrees, 2004, sections 7 and 12; Finnish Ministry of Defence, 2008, §4 and 8; National Defence University of Finland, 2009).

This analysis shows how linearly the objectives of higher learning are set and applied, at least in Finland and its military educational organizations. This seemingly linear process does not mean, however, that there is no interaction when the fruitful military pedagogical meaning of the nationally set aims and objectives is considered. The pedagogical meaning of these aims complements the organizationally defined pedagogical policies and strategies, but does not replace them or diminish their role. In other words, the Bologna-linked principles do not enter, or descend, into a "vacuum". Instead, this "vacuum" is filled with the local cultural aspects, either reinforcing or restricting the applicability of the politically determined Europe-wide principles which are intended to develop our educational institutions.

When I thematically interviewed almost all the senior officers and professors of the FNDC in 2005 (Mäkinen, 2006), the data revealed the cultural fact that the importance of *the comparability* principle had been increased also through the Bologna Process at the FNDC. In this case, the comparability principle meant that the degrees of the FNDC should be comparable with, and on a same academic level as their civilian counterparts. The word "academic" has an ambiguous meaning, and instead of it the respondents emphasized *scientificness*.

According to a majority of them, the biggest reform of the Bologna Process at the FNDC was the emphasis on scientificness. In this case the term "scientificness" was used as a catchword, but nevertheless justifying the collective effort to find out what scientificness means for us in a more analytical and still practical sense (Mäkinen, 2008; Toiskallio & Mäkinen, 2009).

In a contradictory way, a majority of the respondents felt that during the Bologna Process there had been too many and too long discussions slowing down the otherwise rapid educational planning process for the new curriculum of the FNDC. The silencing of organizational discussions could be the most effective way to act if the process was reduced only to the level of the written curriculum. When remembering the other layers of the curriculum, we do not have the option of avoiding discussions, at least if the goal is effective educational development or even educational transformation. If the discussions need a more focused structure, this could be achieved by the identification of the fundamental questions and by starting the planning process from these kinds of questions, instead of decentralizing the issues to be dealt with in the four autonomous academic subjects and eight disciplines being officially represented at the FNDC/FNDU.

The new two-cycle degree system was adopted by Finnish civilian universities in August 2005, and a year later, in 2006, it was adopted by the FNDC. As mentioned above, the FNDC has applied the two-tier degree structure in its system of officer education since 2001. Instead of study weeks, the academic credit points[5] are the Europe-wide reference points for the estimation of the students' workload. The length of the lower academic degree was to remain the same at the FNDC, but the length of the higher academic degree was shortened to two years for most students[6].

In Finland, one of the most important effects of the Bologna process has been the very active process of renewing and updating *the core contents* of disciplinary and professional academic curricula in departments (Jakku-Sihvonen & Niemi, 2006) and in the academic subjects of the FNDC/FNDU as well. Based on their research-based core content analysis, each of the academic subjects, headed by their professors, decided *the course structure* for both BA and MA studies. Therefore the general danger of re-packaging the old content to a new curriculum (Hoffman, Välimaa & Huusko, 2008) has been avoided, because it has been emphasized how strategically important it is that all education is based

---

5   The curriculum consists of two intertwined parts: academic, meaning to be educated and non-academic, meaning to be trained. In this phase it suffices to say that only the academic content of the curriculum is counted with academic study points.
6   The students studying in a parallel manner to be pilots either for the Air Force or the Army are exceptions, as their studies last six years and begin immediately after they have completed their BA studies. Others will work some three to four years before starting studies on a MA course.

on research and professional practices (cf. Finnish Ministry of Education, 2004, section 7; National Defence University of Finland, 2006).

In parallel, when the Bologna Process had begun, the commandant of the FNDC led a planning process for the new strategy of the FNDC. Due to this parallel planning process, the strategy of the FNDC/FNDU was influenced by the ideas and principles of the evolving Bologna Process. But more interestingly, both the strategy itself and the process of its formalization emphasize the centrality of strategic military pedagogic ideas, such as:

- A vision and strategy and other future-oriented considerations for the institution
  - The competitive horizon of the FNDC was challenged and reformulated to be both European and global.
    a) "The National Defence University critically and actively strives to find its role, position, status and special tasks *in the changing field of security, crises, war and defence.*"
  - A vision also comprises some guiding principles[7], shared values[8], and a visionary end state.
    b) "In 2016 the NDU is a nationally and internationally respected university, specialising in security and defence policy as well as national defence policy questions."
- Learners' experiences, expertise and knowledge
  - The main point is not just what the vision is, but how the vision is formed and what the vision achieves. In other words, the only meaningful criteria for judging the vision are the actions and changes that ensue (Senge, 1990; Senge, Scharmer, Jaworski & Flowers, 2005). Therefore, the vision has to be formed in a shared manner; otherwise it cannot be both a personally and collectively shared vision. Consequently the learners, both the students and the teachers, come to the fore.
- Academic subjects in the institution
  - A military academic institution should also do basic research on an ongoing basis. The basic research should focus on the academic subjects and their disciplines (Kesseli, 2007; Mäkinen, 2008; Toiskallio & Mäkinen, 2009). This kind of basic research should be done in both an interdisciplinary and transdisciplinary manner.

---

7 Such as comparability of the degrees, scientificness, alignment, etc. (see Mäkinen, 2006).
8 Such as expertise (see Mäkinen, 2006).

The Bolognian educational planning process was influenced by the parallel process of planning a new strategy for the FNDC. In the Bolognian educational planning process, some additional aspects were emphasized, such as:

- Core content analysis
  - The curriculum consists of both academic and non-academic content, meaning that the officers will be educated as well as trained based on the curriculum.
- Modules of the curriculum
  - The taught content will be divided into courses, and the courses can be grouped into modules (e.g., a module for crisis management operations). At the moment, the BA and the MA courses are not divided into modules at the FNDU. On the other hand, the courses of the Senior Staff Officer Course are divided into modules.
- Course structure
  - The taught content is divided into courses belonging either to the basic, intermediate, or advanced (MA) level. The courses are chronologically aligned along the calendar, and consequently *studying paths* are established.
- Learning objectives
  - For each of the courses, specific learning objectives are set, aligned to the aims of the degree, i.e., to the specific position of the course along the studying path.
  - Individual learning objectives are also needed, and they are defined in the Individual Study Plans and are approved by either the teacher or the professor of the academic subject in question. Especially in MA studies this aspects plays a central role.
- Pedagogical and didactic options and decisions are elaborated by progressively inquiring teachers and students
- Evaluation/assessment principles
- Alignment issues

This discussion about the curriculum, its aims, and objectives, about the actions and activities of the teachers and the students, has brought out the idea of *alignment*. Alignment is both experience and research-based concept emphasizing some guiding principles for the actors of educational institutions. In the field of educational studies, John Biggs (2003) has discussed the meaning of alignment. According to him, alignment (or *constructive alignment*), means that the critical components to be aligned and balanced are the curriculum, the teaching methods, the assessment procedures, the climate with the students, and the institutional climate, i.e., the rules and procedures we have to follow (Biggs, 2003; Hakkarainen,

Lonka & Lipponen, 2004; Lindblom-Ylänne & Nevgi, 2003; Mäkinen, 2006). A slightly simplified explanation of alignment has been proposed by Anderson, Kratwohl, Airasian et al. (2001), to whom alignment means ensuring that instruction and assessment are aligned with the objectives (i.e., the aims). On the other hand, alignment has also other meanings, such as *organizational alignment*, referring to the decisions, behaviors, and acts of the individual actors being aligned with the strategy of the organization, or at least with each other (Mäkinen, 2006; 2007).

How will the evolving Bologna Process be applied at the FNDU in the future?

After focusing on the historical case of applying the Bologna principles to the Finnish military education, it is now time to turn the attention to the future of the Bologna Process and to the institutional adaptation process, in this case of Finland. A brief analysis of the Communiqués of the ministerial meetings of the Bologna process was made above, and their military pedagogical meaning was identified. The original Bologna principles, and even more so the principles made during the Bologna follow-up process, are *underutilized* at least in the Finnish case, but maybe elsewhere as well.

When discussing either over- or underutilization, the issue of *reduction* should be discussed. By reduction I mean the danger of reducing the Bologna Process to be understood as being just about the shared competence language. Instead of just shared languages, definitions, and descriptions, the Bologna Process is, and should be, about shared visions, frameworks, principles, and aims.

As even a brief analysis of the Bologna Process shows, both international and national frameworks of reference were a long time in the coming, at least since the ministerial meeting in Prague in 2001. Finally in 2008, the European Parliament and Council adopted the European Qualifications Framework (EQF). EQF encourages countries to connect their qualification systems or frameworks to EQF by 2010 and to ensure that all new qualifications issued from 2012 carry a reference to the appropriate EQF level (European Commission, 2009).

The core of EQF are eight reference levels describing what a learner knows, understands, and is able to do – "learning outcomes". The levels of national qualifications (i.e., degrees) will be placed at one of the central reference levels, ranging from basic (level 1) to advanced (level 8) (ibid., annex II). This means that for example the BA degrees are on level 6 and MA degrees on level 7. According to EQF, "learning outcomes" (ibid., annex I) mean statements of what a learner knows, understands, and is able to do on completion of the learning process. In other words, the learning outcome is defined in terms of *knowledge, skills* and *competence.*

In 2008, the Finnish Ministry of Education appointed a committee to prepare a national qualification framework (NQF) describing qualifications and other learning. The task of the committee was to prepare a proposal on the national qualification framework and to define its levels in terms of *knowledge, skills* and *competences,* to determine the criteria according to which the qualifications are placed on the different levels of the national and European qualification frameworks, and to propose which levels the Finnish qualifications should be placed in the frameworks. The committee's key proposals are that the Finnish qualification framework will have eight levels based on EQF. The national framework describes the requirements of Finnish qualifications (learning outcomes) in terms of knowledge, skills, and competence, which are the criteria agreed upon in European cooperation based on the EQF levels. But interestingly, at least at this phase, the dimensions of learning are not distinguished from one another, and the EQF levels are specified according to a national perspective. Despite this statement of competence (*osaaminen* in Finnish) orientation, the specific requirements are set in terms of knowledge, skills, and *responsibility* (cf. competence) for the academic degrees (Committee for the Preparation of the National Qualifications Framework, 2009, appendix 1).

The committee's other proposals are worth focusing on, due to their additional military pedagogical meaning. Appendix 1 of NQF claims that on level 7 (MA) the students should have "critical awareness of knowledge issues in a field and at the interface between different fields". On level 6 (BA) there is some difference between the EQF and NQF descriptions. EQF emphasizes that the students should get "advanced knowledge of a field of work or study, involving a critical understanding of theories and principles", whereas NQF claims that the students should be "aware of the boundaries between "practical"[9] and "theoretical"[10] knowledge".

In a way, on this level NQF also emphasizes a holistic view of the field divided into institutional areas where both military science and "civilian sciences" become a space on maps of culture (Gieryn, 1999; Mäkinen, 2006; Toiskallio & Mäkinen, 2009), as illustrated in Figure 1. The boundaries between "civilian science", military science and the ordinary activities of professional officers will be negotiated and renegotiated both individually and collectively on a continuous basis, and the results of these kinds of negotiations will shape the content of the curriculum. One of the emerging fundamental questions is how the experience-

---

9   By explicitly referring to the "field of work".
10 By explicitly referring to the "field of (academic) study".

based competence (i.e., all kinds of prior learning) gained through former studies at schools, workplaces, and elsewhere will be incorporated into NQF and consequently the Finnish curricula (Committee for the Preparation of the National Qualifications Framework, 2009). Therefore, the Europe-wide mechanisms for recognizing both non-formal and informal learning in the evaluation of the students' learning outcomes will play a pivotal role in the future (European Centre for the Development of Vocational Training, 2010). Interestingly, also the formal learning at schools should be included in the scheme to avoid the emerging dangers of lowering the intellectual and practice-oriented requirements of our academic studies.

Figure 2. Cultural boundaries of the curriculum of the officer education

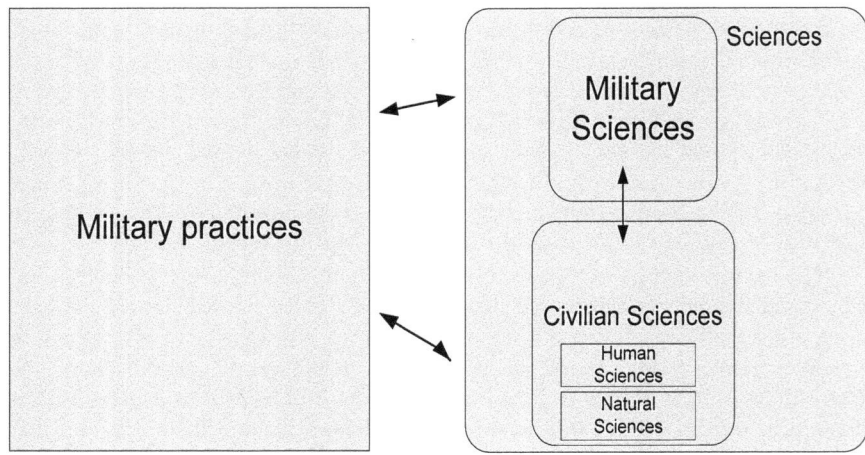

Both EQF and NQF deal with the skills of soldiers. Therefore it is interesting to note that it has traditionally been claimed that military pedagogy is a doctrine of training skills. Military pedagogy has been seen as a doctrine of setting goals, guiding learning, and assessing training activities and know-how (Mäkinen, 2006; Nissinen, 2001; Toiskallio, 1998). Both EQF and NQF challenge this kind of overemphasis on skills and practical knowledge. According to both quality frameworks, not only skills but also knowledge and competences (i.e., action competence, see Toiskallio & Mäkinen, 2009; also generic competences, see Kallioinen, 2008) are needed. Finally, it suffices to emphasize that the national interpretation contradicts not only the international (EQF), but also global academic interpretations (Mäkinen, 2006; 2007; 2008).

In line with Rylean thinking (Mäkinen, 2006), acting intelligently does not embody two processes, one of doing (knowing *how*) and another of theorizing (knowing *that*), but just one. Interestingly, in the process of acting intelligently, both types of knowing are intertwined, being complementary to each other. But what does this academic fact mean on the level of a curriculum?

A Finnish case will explain my point in this issue. In Finland, the content included in the curriculum is divided into two *intertwined* parts of academic and non-academic studies. When the workload of students has remained justifiably a Europe-wide issue[11] (including at the FNDU), one of the main challenges has been the too high workload, especially on some lines of the studies. When the time period allocated for the degree is kept constant (i.e., in principle three years for BA or two years for MA) for all students and the studies are both academic and non-academic, only the workload caused by the academic studies is intended to be estimated and measured by using academic credit points as shared reference points for the workload. So far the non-academic workload has not been officially estimated and measured by any means[12], and therefore the debate on the overload of the studies is an organizational issue.

At least in the FDF, and in the armed forces in general, the constant flow of graduating officers seems to be the main issue; not their real learning outcomes, real workloads, and balanced interpretations of the relations between academic and non-academic studies. In a sense, the traditional chain of graduation ceremonies and graduating officers overrides the need to put the highest emphasis on the learning outcomes, the individual and collective studying processes, and the consequent total workload of the students. In other words, the time period for the studies should not be a constant. Each student should be graduated not before, but after achieving the course-specific learning outcomes in all their studies *at least on the minimum level*. The organizational emphasis on a minimum level means that most of the students will achieve learning outcomes far above the minimum level. This means that the general emphasis on the predefined learning outcomes (i.e., products) should be balanced by the emphasis on the individual and collective learning *processes* for the life-long path of *on-the-job education*[13] and training.

---

11   Interestingly emphasized, e.g., in the Berlin Communiqué in 2003.
12   Informally and unofficially, the individual workload gets measured, and also in the military pedagogical studies students keep learning diaries explaining the experienced reality in a written form, to be utilized in the organizational development processes. At the Stockholm conference on the European initiative for the exchange of young officers (November 25th-27th, 2009) the idea of "military ECTS" (MECTS) was introduced as a shared reference point for e.g., workload of the non-academic studies.
13   Traditionally the term has been "on-the-job training", but there exists a difference between education and training

When the focus remains on the global actions and practices (Giddens & Himanen, 2006; Himanen, 2007; Mäkinen, 2007), we have to keep in mind that instead of a shared competence language, other kinds of on-going objects of our shared attention need to be elaborated further on at the FNDU:

- The strategy of the FNDU and its military pedagogical meaning for the teachers and students in all departments
- The academic subjects are under an on-going institutional reflection, and new links between them should be identified and established by interdisciplinary basic research
- Not only core content analysis is needed, but also interdisciplinary content analysis linking the core contents of the academic subjects to the core content of the degree studies
- The form of the curriculum is optional, and modules are already used in the FNDU
- The learning objectives of the courses can be modified with the European-wide competence language in terms of knowledge, skills, and competences
- The teachers, and also the students, make many pedagogical and didactic decisions on an ongoing basis, and these decisions should be both organizationally and constructively aligned to each other; and not only on paper!

Bibliography

Anderson, L., Kratwohl, D., Airasian, P. et al. (Eds.). (2001). *A Taxonomy for Learning, Teaching, and Assessing: A Revision of Bloom's Taxonomy of Educational Objectives, Abridged Edition.* New York: Addison Wesley Longman.

Benelux Bologna Secretariat (2009). *About the Bologna Process.* Retrieved from http://www.ond.vlaanderen.be/hogeronderwijs/Bologna/about/index.htm.

Biggs, J. (2003). *Teaching for Quality Learning at University.* Buckingham: The Society for Research into Higher Education & Open University Press.

Committee for the Preparation of the National Qualifications Framework (2009). *The national framework for qualifications and other learning.* Helsinki: Helsinki University Print. Retrieved from http://www.minedu.fi/export/sites/default/OPM/Julkaisut/2009/liitteet/tr24.pdf.

European Centre for the Development of Vocational Training (2010). *European guidelines for validating non-formal and informal learning.* Luxembourg: Office for Official Publications of the European Communities. Retrieved from http://www.cedefop.europa.eu/EN/Files/4054_en.pdf.

European Commission (2009). *The European Qualifications Framework*. Retrieved from http://ec.europa.eu/education/lifelong-learning-policy/doc44_en.htm.

European Ministers of Education (1999). *Bologna declaration*. Retrieved from http://www.bologna-bergen2005.no/Docs/00-Main_doc/990719BOLOGNA_DECLARATION.PDF.

European Ministers of Education (2001). *Towards the European Higher Education Area – Prague Communiqué*. Retrieved from http://www.bologna-bergen2005.no/Docs/00-Main_doc/010519PRAGUE_COMMUNIQUE.PDF.

European Ministers of Education (2003). *Realising the European Higher Education Area – Berlin Communiqué*. Retrieved from http://www.bologna-berlin2003.de/pdf/Communique1.pdf.

Finnish Defence Staff (2002). *Finnish Military Defence: Finland's Security Policy; Organisation of National Defence*. Retrieved from http://www.puolustusvoimat.fi.

Finnish Ministry of Defence (2008). *Act on the National Defence University (1121/2008)*. Retrieved from http://www.finlex.fi/fi/laki/alkup/2008/20081121.

Finnish Ministry of Education (2004). *Decree on University Degrees (794/2004)*. Retrieved from http://www.finlex.fi/en/laki/kaannokset/2004/en20040794.pdf.

Giddens. A. & Himanen. P. (2006, June 4th). *Eurooppalainen unelma* [A European Dream]. Helsingin Sanomat.

Gieryn, T. (1999). *Cultural Boundaries of Science: Credibility on the Line*. Chicago: University of Chicago Press.

Gonzales, J & Wagenaar, R. (2008). *Universities' contribution to the Bologna Process. Tuning Educational Structures in Europe*. Retrieved from http://www.unideusto.org/tuningeu/images/stories/Publications/ENGLISH_BROCHURE_FOR_WEBSITE.pdf.

Government of Finland (2008). *Proposal 62/2008*. Retrieved from http://www.finlex.fi/fi/esitykset/he/2008/20080062.pdf.

Hakkarainen, K., Lonka, K. & Lipponen, L. (2004). *Tutkiva oppiminen: Järki, tunteet ja kulttuuri oppimisen sytyttäjinä* [Progressive inquiry: reason, emotions and culture as inspirers of learning]. Porvoo: WS Bookwell.

Haug, G. & Tauch, C. (1999). *Towards the European higher education area: survey of main reforms from Bologna to Prague*. Retrieved from http://ec.europa.eu/education/erasmus/doc/publ/trendssum.pdf.

Himanen, P. (2007). *Suomalainen unelma: innovaatioraportti. Toinen, korjattu pianos* [A Finnish Dream: An Innovation Report]. Helsinki: Artprint.

Hoffman, D., Välimaa, J. & Huusko, M. (2008). The Bologna Process in Academic Basic Units: Finnish Universities and Competitive Horizons. In J. Välimaa & O.-H. Ylijoki (Eds.), *Cultural Perspectives on Higher Education* (pp. 227-243). Netherlands: Springer.

Jakku-Sihvonen, R. & Niemi, H. (2006). The Bologna Process and its implementation in teacher education. In R. Jakku-Sihvonen & H. Niemi (Eds.), *Research-based Teacher Education in Finland – reflections by Finnish Teacher Educators. Research Report 25* (pp. 17-29). Turku: Finnish Educational Research Association.

Joint Quality Initiative (2009). *Agenda*. Retrieved from http://www.jointquality.nl.

Kallioinen, O. (2008). *Generic competences in producing expertise in Military Academy – Case Master of Military Sciences*. Paper presented at the 9th International Conference on Military Pedagogy, National Defence University, Finland.

Kesseli, P. (2007). War Studies at the Finnish National Defence University. In T. Kristiansen & J. Olsen (Eds.), *War Studies: Perspectives from the Baltic and Nordic War Colleges* (pp. 63-85). Norway: Norwegian Institute for Defence Studies.

Lindblom-Ylänne, N. (Ed.). (2003). *Yliopisto- ja korkeakouluopettajan käsikirja* [Handbook of University Teachers]. Vantaa: WSOY.

Lourtie, P. (2001). *Furthering the Bologna Process. Report to the Ministers of Education of the signatory countries*. Retrieved from http://www.ond.vla anderen.be/hogeronderwijs/bologna/documents/GRP_for_MC/LourtieReport-FromBolognatoPragueMay2001.pdf.

Mäkinen, J. (2005). Military education in the age of the Bologna Process. In T. Jaskari (Ed.), *Tiede ja Ase, Vol. 63.* (pp. 161-175). Jyväskylä: Gummerus Kirjapaino.

Mäkinen, J. (2006). *The Learning and Knowledge Creating School: Case of the Finnish National Defence College*. Helsinki: Edita Prima.

Mäkinen, J. (2007). Interplay Between the Culture of the FDF and the Knowledge Management Field. *International Journal of Knowledge, Culture and Change Management, 7*(6), 75-85.

Mäkinen, J. (2008). Military sciences emerging from the intersection of military and civilian cultural spheres. In A. Mutanen (Ed.), *The Many Faces of Military Studies: A Search For Fundamental Questions* (pp. 95-106). Helsinki: Edita Prima.

National Defence University of Finland (2006). *The Strategy*. Retrieved from http://www.mpkk.fi/fi/esittely/strategia.

National Defence University of Finland (2009). *Student Manual 2009-2010*. Helsinki: Author.

Nissinen, V. (2001). *Military Leadership: a critical constructivist approach to conceptualizing, modeling and measuring military leadership in the Finnish Defence Forces*. Helsinki: Finnish Defence College, Department of Management and Leadership.

Paile, S. (2008, November). *The Bologna Process in the Education of the Military Officers: The Road to a Common Culture of Defence in Europe*. Communication at the French EU Presidency Seminar on the Initiative for the Exchange of Young Officers.

Senge, P. (1990). *The Fifth Discipline: Art and Practice of the Learning Organization*. New York: Currency Doubleday.

Senge, P., Scharmer, C., Jaworski, J. & Flowers, B. (2005). *Presence: Exploring Profound Change in People, Organizations, and Society*. New York: Currency Doubleday.

Toiskallio, J. (1998). *Sotilaspedagogian perusteet* [The Basics of the Military Pedagogy]. Hämeenlinna: Karisto.

Toiskallio, J. & Mäkinen, J. (2009). *Sotilaspedagogiikka: sotiluuden ja toimintakyvyn teoriaa ja käytäntöä* [Military Pedagogy: Theories and practices of the soldiership and action competence]. Helsinki: Edita Prima.

Tomusk, V. (2007). Introduction: COM(91) 349 final and the Peripheries of European Higher Education. In V. Tomusk (Ed.), *Creating the European Area of Higher Education: Voices from the Periphery* (pp. 1-18). Netherlands: Springer.

Välimaa, J., Hoffmann, D. & Huusko, M. (2007). The Bologna Process in Finland: Perspectives from the Basic Units. In V. Tomusk (Ed.), *Creating the European Area of Higher Education: Voices from the Periphery* (pp. 43-68). Netherlands: Springer.

Sylvain Paile

Towards a European Understanding of Academic Education of the Military Officers?

Introduction

Giving keys for understanding and acting on the surrounding defense challenges is the main objective pursued by academic education of military officers. In the past two decades, European military education systems had to adapt to global changes of their environment. They made the choice of balancing their curricula for excellence in both military and academic aspects of their education.

On the one hand, new missions have appeared on the scene and changed the relationship between military elites and the society. A new military officer emerged from the post-Cold War era while some *operations other than war* (OOTW) entered his core competencies. Multilateralism, or at least international responsibility for peace maintaining, proved that military officers were meant to serve also supranational purposes. In his/her professional career he/she may now meet civilian participation to his/her mission, and use civilian means. Within the specific European context, military officers are called to deal with the growing integration of the European Security and Defence Policy (ESDP). The ESDP – renamed Common Security and Defence Policy (CSDP) in the Lisbon Treaty – is an ever-evolving policy dealing with 27 individual defense policies nowadays. It may thus be challenged by being considered as a too soft instrument for an effective and efficient European defence. Fashion is to talk currently more about "European security" than "European defence". The former outlines indeed two guiding-principles of the ESDP: "solidarity" and "security".

The 1993 defined Petersberg missions are now broadly assimilated by the European society. Political mechanisms for action were created to conduct this policy but, once again, new missions seem to integrate the list. In that sense, we shall specifically keep an eye on the development of the collective defense mechanism that appeared in Article 42 §7 of the consolidated version of the Treaty on European Union. This legal clause might indeed change both the relationship of Europe with the rest of the world and between the Member States themselves[1]. The military officer must deal with this specific environment as a "soldier-statesman" (Moskos & Burk, cited by Manigart, 2003).

---

[1] This particular article was a major reason for the negative referendum hold in Ireland about the ratification of the Lisbon Treaty, in June 2008.

In parallel, the social environment of the military profession in Europe also faced changes. Transformation of the armed forces themselves led to consider academic education as an instrument for answering these new challenges. Since the mid-1990s, the transition from conscription to professional armies contributed to the loss of brilliant minds for the service of the national armies. Terms of careers became shorter and perspectives of post-service re-conversion turned to be more and more difficult for those short-term officers. In the post-Cold War era we entered in 1991, and with regard to the growing importance of reconstruction missions, social acceptability of the holders of the defense knowledge was at stake. Institutions responsible for the education of the future officers reasonably opted for a more and more academic-oriented education, in order to reverse this "brain drain". The military officer was called to become a "scholar-soldier" (ibid.).

But did the national military officer education systems uniformly adapt to such changes of their environment? An indication of this adaptation has been suggested to be visible in the convergence of military higher education into a civilian-like higher education, similar to a university model. This conception seems to be strongly confirmed by the reforms initiated by schools, academies, and defense universities in their education policies. However, some challenges remain and call for a European education of European defence.

Doctrines for understanding the military education systems

In the 2003 study of Kirkels, Klinkert and Moelker (2003), entitled "*Officer Education: The road to Athens!*", a distinction was made between different traditions of academic education in the context of military officers education. Two models were emphasized in consideration of the nature of the officer and his role in the peace construction: the Sparta model and the Athens model.

The former outlines the need for a military officer to be firstly a soldier with regard to his behavior on the operation field. The latter favors the vision of the military officer belonging to an intellectual elite, ready for dealing with the complexity of the social, economical and policy-related tasks of his/her mission. The authors themselves advocate a more academic-oriented education for military officers, i.e., an Athenian model, relying on five important assessments:

- Focusing the education on combat training remains necessary, but it is not sufficient anymore.
- An education system essentially focused on the teaching of human values and practical knowledge, as it is given in classical academies, might attract an unsuitable clientele with regard to the new missions. Then, it might not be adapted enough to the political demands that are emerging from a European environment.

- The job profile of an officer (i.e., his/her competencies) should correspond better to professional capacities than practical knowledge.
- A growing integration of the national civilian higher education and the military education system is better suited regarding mission flexibility and allows budgetary coherency.
- Military education should follow the universities' standards and, in order to provide an adapted study environment, shall be provided within civilian universities.

A very similar distinction had been made by Giuseppe Caforio (2000) in his study *"The European Officer: A Comparative View on Selection and Education"*. He outlined two models of socialization in military higher education institutions: the divergent model and the convergent model. Caforio examined whether such an institution was following the civilian university model we encounter in most European countries, i.e., convergent with the civilian system, or not, thus being divergent from the civilian system. Six criteria were used to ascertain the socialization process within these institutions:

- Selection procedure;
- Teaching staff;
- Share of academic and military education in the curriculum;
- Chronological organization of these two aspects of the education;
- Civilian value of the diploma;
- Type of socialization favored within the institution.

A classification on a scale (divergent to convergent) was then established.

These two studies neither replicate, nor do they contradict each other. Their objects are different: nature of the officer on the one hand, socialization process on the other hand. However, we shall suppose that an Athenian vision of the military officer requires from individuals a level of excellence in non-exclusive military matters. We may also suppose that a convergent system of education is the best way to attain the open-mindedness that is required from an Athenian officer. Likewise, a divergent system being mostly focused on the military specificity of the professional outcomes would obviously be the most adequate instrument for educating Spartan officers. The two studies were thus calling to increase the importance of academic teaching in officers' education.

The education policies followed by European Professional Military Education Institutions (EPMEIs) is looking forward to reach a conception of the officer, Athenian or Spartan, according to one of the mentioned educational "spirits", convergent or divergent. Is it possible, however, to take the "convergence-divergence" distinction as the only existing instrument for the analysis of the armed forces education policies?

These policies are the instruments that organize this socialization, but they may override the limits of socialization. Education policies indeed may hide their divergent tendencies. Education of a long-term military officer, viewed holistically, is and will necessarily remain shared between the need for accomplished soldiers and the need for global defense actors. The analysis of the overall education of a career officer has to take into account both of these dimensions: the military and the academic. Otherwise, the study of one institution alone may mask the other dimension of the apprenticeship, even if it presents all the characteristics of a civilian university. The existing differences observed between educational models in Europe seem thus slightly minor than at first sight[2], and we shall certainly enlarge the study sample from the institution to the overall educational organization.

Nevertheless, the criteria set by Caforio's study remains essential for the search for exchanges and cooperation and, furthermore, for the study of national defense cultures. They are of major importance to find out what "spirit" lies behind a particular organization of curriculum. This distinction remains very accurate also for monolithic education systems, i.e., institutions in charge of the entire education of the officer, both military and academic.

Two paths may be taken to attain a convergent model by means of education policies:

1. Increase the importance of academic education in a curriculum;
2. Delegate the academic part of the officer education to the civilian higher education system.

The latter case is handled differently in many of the European military education systems. We have chosen three cases to illustrate the importance of this delegation: the British system, the Slovenian system, and the French system.

In the British system, according to Caforio's criteria, the Sandhurst Academy might appear as being a divergent institution. Nevertheless, studies have shown that on average, cadets have already reached an important academic level in civilian universities prior to their recruitment (Alber, 2007).

In the Slovenian system, military officers are recruited from the poll of master students who have graduated from civilian universities. Most of them actually come from the Faculty of Social Sciences of the University of Ljubljana. The military training starts only after their recruitment, within specialized schools. The Slovenian case suggests an intentional delegation of the academic training to

---

[2] If we were to take the German system of education as an example, the socialisation process within the universities of the Bundeswehr would appear as being obviously convergent, although the military aspect of education is "delegated" to other military education institutions.

the civilian higher education system. It is thus equally uneasy to qualify it as a divergent or convergent education.

In the French system, two possibilities exist concerning recruitment: cadets can integrate military schools after having obtained a university degree or after having followed two academic years within "preparatory classes". The first option signifies that the French Ministry of Defence has no power over the pre-socialization process of cadets. In the second option, it is involved in the pedagogical concerns due to the *raison d'être* of these preparatory classes.

A third path for the recruitment of French officers is called recruitment "*sur titre*". It allows master-graduated students from civilian universities to integrate officer school. In that situation, cadets would only accomplish a military training curriculum. It is then quite uneasy to define the Saint-Cyr Schools as being divergent or convergent, following the particular options.

Eight years after his publication, new kinds of missions have integrated the competencies of European armed forces in a way confirming the trends on which Caforio based his distinction. The European Union got a Strategy in 2003 that includes the need for positive actions, such as reconstruction of peace or nation building. These missions enhance the need for flexible and more autonomous officers, and officers able to react with regard to a certain understanding of the global situations. The academic aspect of officer education has effectively been growing in importance due to reforms initiated by the educational systems during the last decade.

Post-reforms observations of few European educational models

The implementation of the Bologna Process, even if slightly different from one educational system to another, confirmed the trends toward a growing need for academic education. The development of programs and exchanges between the different institutions – sometimes using the same instruments as their civilian counterparts (e.g., the "Erasmus" programs) – actively contributed to this "Europeanization" of the educational challenges. The Bologna Process, initiated with the June 19, 1999 Common Declaration of the European Education Ministries – implementing the European Cultural Convention of the Council of Europe – is meant to remove obstacles to mobility in higher education. Three main directions are outlined in this text for attaining this particular objective:

- Diploma recognition;
- Curricula harmonization;
- Quality-assurance of the education.

The signature of the now 47 member states of the Bologna Process has no constraining effect, though. The goal is not a proper "harmonization", but a convergence of the education systems on a voluntary basis. The Process is not linked to supranational legal sanctions in case of delay in the implementation. Some differences may thus be observed between the member states.[3]

The recognition of the three university cycles is set with the implementation of the European Credit Transfer System (ECTS) in the curricula. Practically, an ECTS credit represents an amount of student workload of 25 to 30 hours, depending on the country (European Commission, 2012). According to the evolution of the prescriptions of the Process itself, outcomes expected from an educational module shall be taken into consideration when assigning ECTS. The duration of the cycle is in no way involved in the mutual recognition of the value of this education. Only credits attached are important for the Process' implementation: 180 ECTS minimum for the first cycle (Bachelor), 90 to 120 for the second (Master). Doctorate level education is left to national coordination.

A monitoring is provided by a rotating presidency assisted by a secretariat composed of officials of higher education ministries since 2005. The presidency collects data concerning the implementation of the Process and presents a stocktaking report during the presidency final meeting in which it suggests complementary actions to be taken in order to promote and develop the European Higher Education Area (EHEA). In 2005 in Bergen, Norway, for example, the conference issued a document on quality-assurance standards in education (ENQA, 2005), which led eventually to the creation of register of European quality assurance agencies (ENQA, 2007).

Military higher education is a very particular island in the Process and its implementation suffers somehow from structural specificities related to the European Professional Military Education Institution's (EPMEI) missions. Formally, stocktaking procedures exclusively take into account data communicated by higher education ministries, which are the main actors concerned by these challenges. Consequently, curricula administratively depending on other ministries are little or not considered in these data. Basic military officers' curricula, which are generally under the authority of the Ministries of Defence, are also in that category. Nevertheless, instruments have been created in order to assist some of these higher education institutions in their implementation of the Bologna Process, as for example the "Tuning" program for art and music curricula, which in many countries are under the authority of Culture Ministries (Gonzales & Wagenaar, 2008). However, these curricula are not included in the data collected for the monitoring of the Bologna Process and no such program has been planned yet to include the basic military officer curricula. It is thus only on a voluntary basis that the systems might have integrated these prescriptions.

---

3 A report on the national progresses towards implementation of the Process is made by the Presidency and presented at the biannual conference. It also suggests means for attaining the planned objectives. Cf. the Bologna Process Stocktaking Report (2007).

The following charts present an overview of the Bologna forms of implementation met in some EPMEIs. They are taken from Paile (2009), a publication made from the report of the research for the Belgian Federal Science Policy (SPF) "Adaptation of the education policy for the Royal Military Academy to the evolution of the Common European Security and Defence Policy" (April 2007- March 2008) which concluded a research program issued by the Royal Military School of Belgium and was coordinated by the Federal Scientific Policy of Belgium. These charts represent the land forces officer education process in a limited number of European countries[4]. Their education systems were studied in order to evaluate the potentialities of exchanges between some institutions of these countries and the Royal Military School of Belgium.

A wide variety of implementation forms may be observed from these figures. In a sense it confirms that the Bologna Process does not intend to harmonize the curricula. However, regarding the purpose of mobility stressed in the course of the Process itself, these differences might potentially obstruct effective exchanges between EPMEIs. The main observation concerns the organization of the curricula themselves. It is indeed necessary for the EPMEIs to reconcile scientific and military training. Schedules differ from one system to the other. While some cadets might be training in camps, others might be pursuing their academic studies. Difficulties might also occur when formulating the learning outcomes expected from a study period. In the same way, academic periods do not follow the same organization from one EPMEI to another. The universities of the Bundeswehr follow a trimester organization while others follow semester organization. A German cadet studying at a different EPMEI would thus disrupt the course of his proper curriculum at home if he decided to attend a whole educational unit abroad.

Finally, these structural observations should not overshadow the fact that these internal Bologna reforms contributed to the growth of academic education in the basic officer curricula compared to the last decades. Even before the start of the Process, Caforio (2000) foresaw this evolution: "The trend towards a convergence of officers' basic educational processes with university appears generalized in Europe and destined to continue, at least in the medium term" (p. 15). The Bologna Process implementation by these EPMEIs tends to confirm his hope. Even if slightly different from one system to another, the space awarded to academic knowledge will certainly contribute to the *entreprise d'excellence* pursued by modern officer higher education.

---

[4] The letter "n" in the following charts represents the start of the curricula, generally the year after the end of secondary education.

Figure 1. Organisation of the academic education in military officers' national education systems

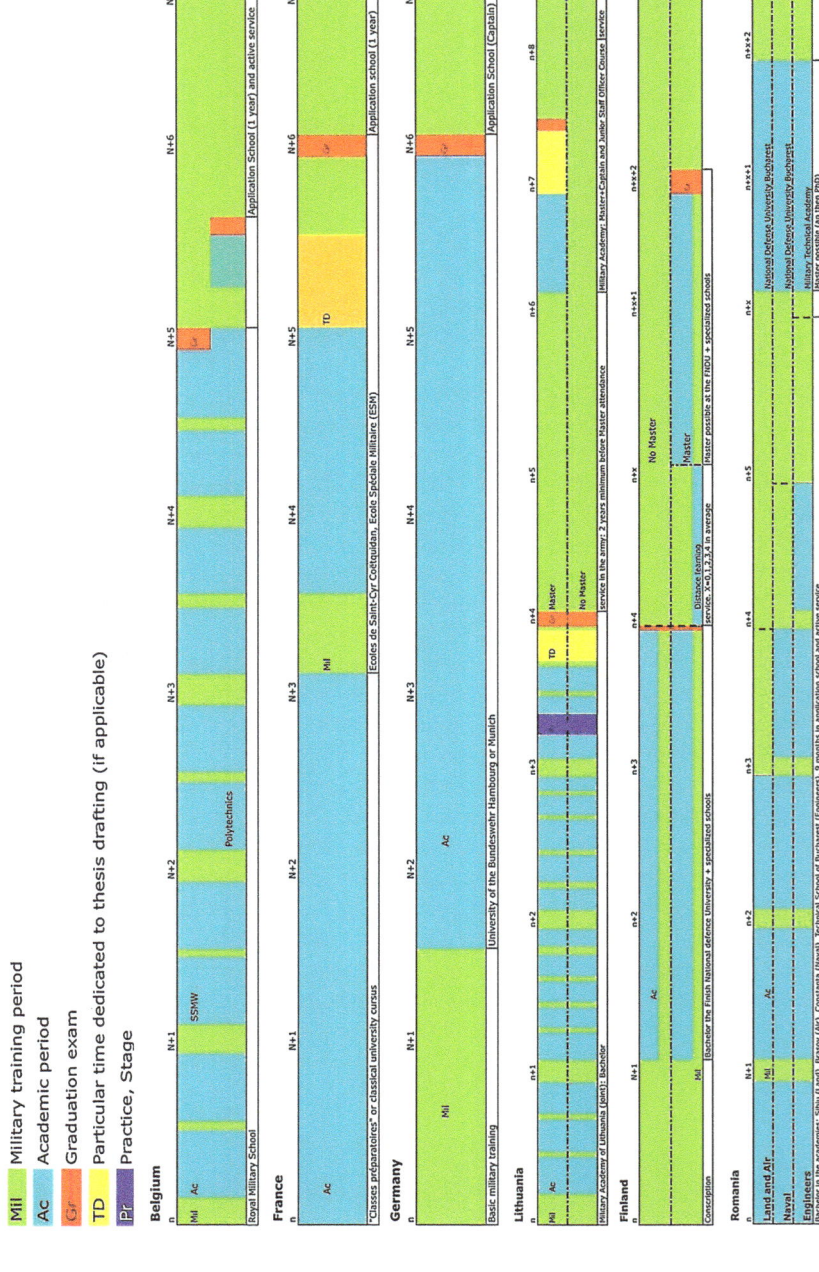

286

Prospective view on tools of a European education for a European Defence

What does the Bologna Process mean with regard to the cooperation between the European military officers education systems? Concretely, it is meant to allow institutions to exchange students and teaching staff on an equality basis. The institutions that have implemented the Bologna Process might also exchange teaching modules thanks to the ECTS system. Recognition of teaching values is made in attaching ECTS credits to courses given within an institution. This allows a student to meet the required number of credits for an academic year by attending courses in a partner institution of his national school.

However, the Bologna Process does not solve all of the practical problems linked to mobility. As we mentioned earlier, it did not prescribe any schedule organization for the curricula. As explained, a student from a school following a semester system may experience schedule problems if he is hosted as an exchange student by an institution with a trimester system. So when would be the most suitable period for exchanges in the curriculum? The Bachelor is both the first socialization period for a cadet with his future professional environment and an apprenticeship of the national dimension of defense and the required behavior as a future military officer. In a number of national education systems it is also a necessary period required for improving the foreign language skills of a young student. The role of language will undoubtedly reveal itself as fundamental in the challenges facing mobility. For these reasons, it appears that the most adequate time frame for actual exchanges between EPMEIs is the master level education. It is important not to exclude from the search for potential partners institutions which provide master courses on the advanced officer level, as in case of Lithuania or Romania[5]. In that sense, the Bologna's mobility purpose would somehow challenge the distinction between basic education and advanced education of officers. In that situation a major issue would be to cope with the social and professional experience differences between the students taking part of the same education.

What could be the future of the integration of the education models into a European context? In the broad majority of national military officers' curricula there is a movement towards convergence of the socialization processes with the civilian higher education systems. Furthermore, and linking it to Kirkels, Klinkert and Moelker's (2003) distinction, the trend is to see the European officer in the context of the European Security and Defence Policy as an Athenian. These

---

[5] See charts presented above. Finnish masters courses are difficult to characterise due to the possibility offered to attend them shortly after the bachelor. Career experience is thus not comparable with Lithuanian or Romanian master officers.

tendencies are the sign of the emergence of a European culture in the military officer curricula. We ought to go further by striving to implement a culture that is distinctive of European officers and directly linked to the future of their missions: an "ESDP culture". The objective would not be to force this dynamic, but rather ease its implementation in the educational systems with help of the EPMEIs education policies, which are direct instruments for action.

In this sense, we may propose to distinguish three paths for acting on the ESDP acculturation: a formal direction, a normative one, and a "crossover" possibility.

The *formal* direction would accustom students to the role they might fulfill in the European defense context. While the European armed forces may be involved in diverse forms of multinational operations such as United-Nations operations or NATO missions, flexibility shall be outlined as a major asset of the European officer's nature. At the institutional level, that would notably imply a debate within the EPMEIs on the conditions of use of languages such as English, or on the use of cooperation instruments such as Erasmus.

The *normative* direction deals with the students' knowledge related to the European defense issues. Inasmuch, this pillar is subject of debate only on the level of the national EPMEIs, e.g., number of courses related to such issues, importance of these teachings in the curricula (compulsory or voluntary), ECTS attached, schedules, etc.).

The *"crossover"* possibility combines both the formal and normative directions in providing an adequate European environment to a specific ESDP education. A practical implementation of this idea might be developed through projects of combined education. The organization of common academic modules by partner institutions would provide the students with a common knowledge in a common environment. This would allow academic resources to be rationalized and common values may arise from these social interactions. The initiative for a "Military Erasmus" currently launched at the European level and coordinated by the French EU Presidency will certainly also concentrate on this crossover dimension.

Conclusions

The perfect "convergence" with civilian university socialization is not meant to exist with regard to officers' higher education. Caforio himself obviously did not mean it either in choosing his terminology: "convergence" is the representation, the idea, of a dynamic, not of a finishing line. The idea of convergence itself has slightly changed when comparing it to the time it was defined, as it is actually *"en route"* in most of the national models. The officer's higher education is becoming an intellectual process. Less and less people may be found that would challenge this new status quo. Even if we cannot meet one common standard in education forms, we may say today that we can observe an actual convergence *towards* convergence. The implementation of the Bologna Process in 2010 is one of these signs that obviously promote the discussion of an actual mobility of the students and teaching staffs. If the possibility is given to the greatest number of students-officers, these movements will trigger an acculturation that will improve their role as members of the society, as students, and as future actors of the ESDP. Efforts must be made now in fostering these expectations by building the relevant instruments for these particular students, who will be battlefield deciders and intellectual actors.

Bibliography

Alber, A. (2007). *La formation initiale des officiers: Une comparaison européenne.* Unpublished PhD thesis, Université de Saint-Quentin-en-Yvelines.

Bologna Process Stocktaking Report (2007). *Report from a working group appointed by the Bologna Follow-up Group to the Ministerial Conference in London.* Retrieved from http://www.ond.vlaanderen.be/hogeronderwijs/bologna/documents/WGR2007/Stocktaking_report2007.pdf.

Caforio, G. (Ed.). (2000). *The European officer: A comparative view on selection and education.* Pisa: Edizioni ETS.

European Association for Quality Assurance in Higher Education (ENQA). (2005). *Standards and Guidelines for Quality Assurance in the European Higher Education Area (ESG).* Retrieved from http://www.ond.vlaanderen.be/hogeronderwijs/bologna/documents/Standards-and-Guidelines-for-QA.pdf.

European Association for Quality Assurance in Higher Education (ENQA). (2007). Report to the London Conference of Ministers on a European Register of Quality Assurance Agencies. *Occasional Papers, 13.* Retrieved from http://www.enqa.eu/files/ENQA%20occasional%20papers%2013.pdf.

European Commission (2012, March 26). *European Credit Transfer and Accumulation System (ECTS).* Retrieved from http://ec.europa.eu/education/lifelong-learning-policy/ects_en.htm.

Gonzales, J & Wagenaar, R. (2008). *Universities' contribution to the Bologna Process. Tuning Educational Structures in Europe.* Retrieved from http://www.unideusto.org/tuningeu/images/stories/Publications/ENGLISH_BROCHURE_FOR_WEBSITE.pdf.

Manigart, P. (2003). Restructuring the Belgian Royal Military Academy. In H. Kirkels, W. Klinkert & R. Moelker (Eds.), *Officer Education: The Road to Athens! NL-Arms* (pp. 121-136). Faculty of Military Sciences of the Netherlands Defence Academy.

Paile, S. (2009). *L'Enseignement militaire à l'épreuve de l'Européanisation: Adaptation de la politique de l'enseignement pour l'Ecole royale militaire de Belgique aux évolutions de la PESD. Thématique du Centre des Sciences Sociales de la Défense, Vol. 19.* Paris: SGA Ministère de la Défense.

The Authors

ANNEN, Hubert. Colonel, PhD, Head of Military Psychology and Military Pedagogy and Dean of the Teaching and Research Branch at the Military Academy at ETH Zurich. Head of the Assessment Centers for career officer candidates, career NCO candidates, and general staff officer candidates. Head of Research of the Psychological-Pedagogic Service of the Swiss Armed Forces. Research in military pedagogy, evaluation and validation of assessment and selection procedures for military leaders, motivational aspects in military context and personality and organizational attitudes.
hubert.annen@milak.ethz.ch

ANTTILA, Ulla. PhD in Military Sciences (National Defence University, Helsinki) and MA in Social Sciences (University of Helsinki). Acting executive director of KIOS, the Finnish NGO Foundation for Human Rights. M.P. from 1991 to 2007. Doctoral dissertation "Enhancing Human Security through Crisis Management – Opportunities and Challenges for Learning" (2012).
ulla.anttila@pp3.inet.fi

CAFORIO, Giuseppe. General (ret.) of the Italian Army. Degree in Strategic Sciences, Law, Political Science, and a Master in Communication Science. Formerly president of the Research Committee 01 "Armed Forces and Conflict Resolution" of the International Sociological Association, vice-president of the Italian Interuniversity Centre of Historical and Military Studies, founding member and coordinator of a working group of the ERGOMAS, and member of several other institutes of research. Outside consultant of the governmental Italian Centre of Strategic and Military Studies.
gcaforio@fastwebnet.it

FOOT, Peter. PhD, Professor Emeritus Defence Studies at the Canadian Forces College, Toronto and the Royal Military College, Kingston, Canada. Former Academic Dean and current Associate Fellow at the Geneva Centre for Security Policy. Research include the issues of religion and violence and the impact of World War II on strategic imagination today. Academic lead on the NATO team visiting military training establishments for the EAPC Defence Education Enhancement Programme (DEEP).
p.foot47@gmail.ch

HUHTINEN, Aki-Mauri. PhD, lieutenant colonel (G.S.), professor, Docent of Practical Philosophy at the University of Helsinki, Docent of Social Consequences of Media and Information Technology at the University of Lapland, and Docent of Information Security and Information Operations at the University of Technology in Tampere. Works in the Department of Leadership and Military Pedagogy at the Finnish National Defence University.
aki_huhtinen@hotmail.com

KASHER, Asa. Laura Schwarz-Kipp Professor Emeritus of Professional Ethics and Philosophy of Practice, Professor Emeritus of Philosophy at Tel Aviv University, and co-author of the first IDF code of ethics. Laureate of Israel Prize for accomplishments in Philosophy.
asakasher@hotmail.com

KRAFT, Luiza. PhD, Associate Professor at the Foreign Language Department (Command and Staff Faculty) of "Carol I" National Defense University, Bucharest, Romania. Director of the Foreign Language Department and teacher trainer. Research in organizational communication, intercultural education, and English language teaching methodology.
luiza_kraft@yahoo.com

MÄKINEN, Juha. PhD in Knowledge Science from the Japan Advanced Institute of Science and Technology, Professor of Military Pedagogy for the Department of Leadership and Military Pedagogy at the Finnish National Defence University. General staff officer in the Finnish Defence Forces prior to his academic career at the National Defence University. Recent publications (in English) include "The Learning and Knowledge Creating School: Case of the Finnish National Defence College" (2006), "Military Pedagogical Comments on the Expeditionary Mindset – A Finnish Interpretation" (2011) (in Furst & Kummel "Core Values and the Expeditionary Mindset: Armed Forces in Metamorphosis"), and "Networked company activities – the case of a Kymi company" (2012) (paper presented at the EGOS conference in Helsinki).
juha.makinen@mil.fi

MÄLKKI, Juha. Lieutenant colonel, PhD in Social Sciences (University of Helsinki). Principal Researcher at the Concept Development and Experimentation Centre of the Finnish Defence Forces. Research in war studies, military sociology and cultural studies.
juha.malkki@mil.fi

MÄLKKI, Kaisu. PhD in Philosophy. Postdoctoral researcher at the Department of Teacher Education, Faculty of Behavioral Science, University of Helsinki. Research in adult learning, reflection and emotions in learning.
kaisu.malkki@helsinki.fi

MALOS, Gavril. PhD in Engineering, colonel in the Romanian Armed Forces, Head of Education and Defence Resources Department, "Carol I" National Defence University, Romania. Teaching and research activity focuses on human resources management and defense resources management.
gabriel_radu2000@yahoo.com

NAKKAS, Can. MSc, Research assistant at the Department of Military Psychology and Military Pedagogy Studies at the Military Academy at ETH Zurich. His primary research interest is the influence of stress on fitness for duty and service motivation.
can.nakkas@milak.ethz.ch

PAILE, Sylvain. Researcher at the European Studies Unit (Department of Political Science) of the University of Liège, Belgium, and expert for the European Security and Defence College (European External Action Service) on the European integration of military education and training.
spaile@ulg.ac.be

PETRAUSKAITE Audrone. PhD, Associate Professor at the Department of Humanities of the General Jonas Zemaitis Military Academy of Lithuania. Research in military ethics, civic education, and military pedagogy.
audrone.petrauskaite@gmail.com

PULKKA, Antti-Tuomas. Major, PhD student (University of Helsinki). Research in military pedagogy, ethics, and motivation in learning at the Department of Leadership and Military Pedagogy of the Finnish National Defence University
antti-tuomas.pulkka@mil.fi

RAVIV, Amira. Degree in Psychology and Master in Organizational Behavior from Tel Aviv University. Lieutenant colonel (ret.), served as the head of the first Learning and Development Center of the Israeli Defense Forces Colleges, and also as the head of the Military Psychologists Ethics Committee. Serves as a strategic consultant for the Ministry of Defense and other security and governmental institutions, and works with international and leading Israeli business organizations. Lecturer in the MBA program of the Faculty of Business Administration. Interests include organizational development, ethics implementation,

learning procedures, counseling and implementation of strategic procedures, and personal coaching to senior leaders.
amira_ra@zahav.net.il

ROYL, Wolfgang. PhD, Professor (retired) at the University of the German Bundeswehr, Munich. Dean for General Didactics at the Faculty of Education since 1976. Member of the German Association for Educational Sciences. Work includes publications on military education written in cooperation with the "Military and Social Sciences" in Strausberg, the German-Atlantic Association in Bonn, and the Hanns-Seidel-Foundation in Brussels.
W.Royl@unibw-muenchen.de or W.Royl@t-online.de

TALERUD, Bo. PhD, Associate Professor, university lecturer and researcher in adult education at the Swedish National Defence College. PhD from Stockholm University, doctoral thesis "Cultural pedagogy in a technological society – On the possibilities of cultural development through pedagogical activities". Current research focus on learning and leadership, learning education and culture, the history of adult education, ethics and the philosophy of education.
bo.talerud@fhs.se

WHETHAM, David. PhD, Senior Lecturer for King's College London, based at the Joint Services Command and Staff College at the UK Defence Academy. Publications include "Ethics, Law and Military Operations" (2010), aimed at practitioners at the operational level of war, the monograph "Just Wars and Moral Victories" (2009) as well as a wide range of journal articles and book chapters on subjects relating to ethics, norms, laws of war, and professional military ethics education.
dwhetham.jscsc@da.mod.uk

Studies for Military Pedagogy, Military Science & Security Policy

Editors:
Dr. H. Jung
Prof. Dr. W. Royl

Band 1 Edmund A. van Trotsenburg: Militärpädagogik. 1989.

Band 2 Hermann Jung / Heinz Florian: Grundlagen der Militärpädagogik. Eine Anleitung zu pädagogisch verantwortetem Handeln. 1994.

Band 3 Franz Kernic: Demokratie und Wehrsystem. Aufsätze zum Verhältnis von Gesellschaft, politischem System und Heer in Österreich. 1997.

Band 4 Edwin R. Micewski: Grenzen der Gewalt – Grenzen der Gewaltlosigkeit. Zur Begründung der Gewaltproblematik im Kontext philosophischer Ethik und politischer Philosophie. 1998.

Band 5 Rudolf Egger / Heinz Florian (Hrsg.): Pädagogische Professionalisierung im Bundesheer. Dokumentation und Reflexion des PädAk-Sonderstudienganges Wehrpädagogisches Management. 1999.

Band 6 Franz Kernic / Harald Haas: Warriors for Peace. A Sociological Study on the Austrian Experience of UN Peacekeeping. 1999.

Band 7 Franz Kernic / Jean M. Callaghan / Philippe Manigart: Public Opinion on European Security and Defense. A Survey of European Trends and Public Attitudes Toward CFSP and ESDP. 2002.

Band 8 Heinz Florian (ed.): Military Pedagogy – An International Survey. 2002.

Band 9 Edwin R. Micewski / Hubert Annen (eds.): Military Ethics in Professional Military Education – Revisited. 2005.

Band 10 Hubert Annen / Wolfgang Royl (eds.): Military Pedagogy in Progress. 2007.

Band 11 Hubert Annen / Wolfgang Royl (eds.): Educational Challenges Regarding Military Action. 2010.

Band 12 Hubert Annen / Juha Mäkinen / Can Nakkas (eds.): Thinking and Acting in Military Pedagogy. 2013.

www.peterlang.de

www.ingramcontent.com/pod-product-compliance
Ingram Content Group UK Ltd.
Pitfield, Milton Keynes, MK11 3LW, UK
UKHW061921210426
5322IPUK00007B/187